Soil Fertility and Nutrient Management

Soil Fertility and Nutrient Management

Editor

Roma Doshi

scitus
academics

Soil Fertility and Nutrient Management

Edited by **Roma Doshi**

ISBN: 978-1-68117-095-4

Library of Congress Control Number: 2015935494

SCITUS Academics LLC,
www.scitusacademics.com
Box No. 4766, 616 Corporate Way,
Suite 2, Valley Cottage,
NY 10989

Printed in United States of America on Acid Free Paper ∞

Contents

Preface

Soil fertility refers to the ability of a soil to supply plant nutrients. Bioavailable phosphorus is the element in soil that is most often lacking. Nitrogen and potassium are also needed in substantial amounts. For this reason these three elements are always identified on a commercial fertilizer analysis. For example a 10-10-15 fertilizer has 10 percent nitrogen. Inorganic fertilizers are generally less expensive and have higher concentrations of nutrients than organic fertilizers. Also, since nitrogen, phosphorus and potassium generally must be in the inorganic forms to be taken up by plants, inorganic fertilizers are generally immediately bioavailable to plants without modification. However, some have criticized the use of inorganic fertilizers, claiming that the water-soluble nitrogen doesn't provide for the long-term needs of the plant and creates water pollution. Slow-release fertilizers may reduce leaching loss of nutrients and may make the nutrients that they provide available over a longer period of time. Soil fertility is a complex process that involves the constant cycling of nutrients between organic and inorganic forms. As plant material and animal wastes decompose they release nutrients to the soil solution. As plant material and animal wastes are decomposed by micro-organisms, they release inorganic nutrients to the soil solution, a process referred to as mineralization. Those nutrients may then undergo further transformations which may be aided or enabled by soil micro-organisms. Like plants, many micro-organisms require or preferentially use inorganic forms of nitrogen, phosphorus or potassium and will compete with plants for these nutrients, tying up the nutrients in microbial biomass, a process often called immobilization.

Editor

Chapter 1

Humic Substances Stimulate Maize Nitrogen Assimilation and Amino Acid Metabolism at Physiological and Molecular Level

Silvia Vaccaro[1], Andrea Ertani[1], Antonio Nebbioso[2], Adele Muscolo[3], Silvia Quaggiotti[1], Alessandro Piccolo[2], and Serenella Nardi[1]

[1]Department of Agronomy, Food, Natural Resources, Animal and Environment, University of Padua, Agripolis - Viale dell'Università, 16, Legnaro 35020, Padua, Italy

[2]Centro Interdipartimentale di Ricerca sulla Risonanza Magnetica Nucleare per l'Ambiente, l'Agro-Alimentare ed i Nuovi Materiali (CERMANU), Università di Napoli Federico II, Via Università 100, Portici 80055, Italy

[3]Agriculture Department, Mediterranea University of Reggio Calabria, Feo di Vito, Reggio Calabria 89060, Italy

ABSTRACT

Background

The effects of a humic substance (HS) extracted from a volcanic soil on the nitrate assimilation pathway of *Zea mays* seedlings were thoroughly examined using physiological and molecular approaches. Plant growth, the amount of soluble proteins and amino acids, as well as the activities of the enzymes involved in nitrogen metabolism and Krebs cycle, were evaluated in response to different HS concentrations (0, 1 and 5 mg C L^{-1}) supplied to maize seedlings for 48 h. To better understand the HS action, the transcript accumulation of selected genes encoding enzymes involved in nitrogen assimilation and Krebs cycle was additionally evaluated in seedlings grown for 2 weeks under nitrogen (N) sufficient condition and N deprivation.

Results

HS at low concentration (1 mg C L^{-1}) positively influenced nitrate metabolism by increasing the content of soluble protein and amino acids synthesis. Furthermore, the activity and transcription of enzymes functioning in N assimilation and Krebs were significantly stimulated.

Conclusions

HS treatment influenced the gene expression of *Zea mays* plants at transcriptional level and this regulation was closely dependent on the availability of nitrate in the growth medium.

BACKGROUND

Humic substances (HS) are the major components of soil organic matter (SOM) and positively affect crop production [1] by playing pivotal roles on both soil-plant system [2] and plant metabolism and development [3]-[6]. The growing concern for sustainable agriculture, whose goals are to ensure lower environmental costs and long-term productivity of agro-ecosystems [7], highlights the importance of developing management strategies to maintain and 1protect soil resources and, in particular, SOM [8]. Despite this present awareness, the second half of the past century has seen a massive and indiscriminate use of high-energy input in agriculture and in particular of N fertilizers during the so-called green revolution. These practices led to the selection of genotypes, which were highly productive when fed with high nitrogen inputs, though not efficient in nitrogen use [9], while they concomitantly caused a progressive impoverishment of soil physical and biological properties, as well as of SOM and nutrients' bioavailability.

Because of the importance of HS in soil fertility, a larger number of works focused on both the chemical-physical structure and the biological properties of humic matter. HS have been traditionally described as polydispersed heterogeneous organic compounds with large molecular weight [10]. However, recent experimental findings reached a new understanding of the humic conformational nature that is regarded as a supramolecular association of heterogeneous molecules with relatively low molecular weight (≤1.5 kD) held together in only apparently large molecular sizes by weak interactions, such as hydrogen and hydrophobic bondings. The metastable conformations of humic associations can be reversibly disrupted by interactions with small amounts of organic acids [11], [1]2], while the same amount of acids do not alter conformation of the true macropolymers stabilized by covalent bonds [13]. The supramolecular nature of humic matter and its response to organic acids have been advocated to explain the bioactivity of humus on plants [6], its molecular dynamics [14] and the slow release of sorbed contaminants [15]. Furthermore, the recognition that NOM is composed by supramolecular associations rather than macropolymers has allowed the development of a fractionation strategy, called humeomics, that enables the analytical detection of most of the single molecules which constitute the supramolecular assembly [16]-[18]. These findings had profound implications on our understanding of SOM function and reactivity.

Several studies have reported the positive effect of HS on crop yield [19]-[21] and on root and shoot development [6], [22]. In addition, leaf chlorophyll content [1], nutrient uptake [3], [23]-[26] and the activities of enzymes involved in several physiological pathways, such as nitrogen assimilation [3], [24], [25], [27] and energy metabolism [28]-[30], seem to be positively affected by HS. However, because of the complexity of HS′ nature, the relationship between their molecular structure and biological activity is not yet completely clear. In the last decades, several studies correlated biological activity to humic chemical features [31]-[41]. The positive effects of different HS in relation to their chemical structure were shown by Muscolo et al. [35] and Nardi et al. [36] on *Pinus nigra* callus metabolism and *Zea mays* seedling growth. Moreover, Zancani et al. [42],[43] successfully related humic matter molecular features to metabolism of tobacco BY-2 suspension cell cultures and to changes of cellular ATP and glucose-6-phosphate during embryogenesis of *Abies cephalonica*.

Furthermore, the complexity of the biological action exerted by HS on plant metabolism suggested the existence of hormonal mechanisms, in particular of an auxin-like effect [3], [4]. The presence of low amount of IAA in different soil humic substances has been reported [33], [44]. Moreover, these substances seem to have a physiological effect on both a maize isoform of H^+ AT Pase *Mha2* [34], [37], [38], which is a specific auxin target, and a pea phospholipase A_2[45], which is a component of an auxin-dependent signalling.

The objective of this work was to investigate the effect of humic matter on maize nitrate assimilation and amino acids metabolism when the humic concentration was varied. The response of growth, protein and free amino acid content of maize seedlings to humic supply was evaluated here. Moreover, the activity of the main enzymes involved in the Krebs cycle and nitrogen metabolism was determined, together with the expression of genes encoding the enzymes involved in these two metabolic pathways.

METHODS

Soil Humic Matter Extraction and Chemical Characterization

A humic acid (HA) from a Fulvudand soil of the volcanic caldera of Vico, near Rome, Italy, was isolated by standard methods as reported elsewhere [46]. The HA was titrated to pH 7.2 with a 0.5 M KOH solution in an automatic titrator (VIT 90 video titrator, Radiometer,

Copenhagen, Denmark) under N_2 atmosphere and stirring. After reaching the constant pH 7.2, the solution containing potassium humate was left under titration for 2 h, filtered through a glass microfibre filter (Whatman GF/C; GE Healthcare, Buckinghamshire, UK) and freeze-dried. The elemental analyses and the molecular characterization of this HA by online thermochemolysis-GC-MS and CPMAS-NMR spectroscopy were reported earlier [36].

Plant Material, Growth Solutions and HS Treatment

Maize seeds (*Z. mays* L. var. DK 585) were soaked in distilled water for one night in running water and germinated for 60 h in the dark at 25°C on a filter paper wet with 1 mM $CaSO_4$[47]. Then, the maize seedlings were raised in hydroponic conditions for 14 days in growth chamber with 450 mL of a complete nutrient solution (μM) containing KH_2PO_4 (40), $Ca(NO_3)_2$ (200), KNO_3 (200), $MgSO_4$(200), FeNaEDTA (10), H_3BO_3 (4.6), $CuCl_2{\cdot}2H_2O$ (0.036), $MnCl_2{\cdot}4H_2O$ (0.9), $ZnCl_2$ (0.09) and $NaMoO_4{\cdot}2H_2O$ (0.01). The solution was renewed every 48 h. Growth chamber conditions were the following: day/night period of 14/10 h, air temperature of 27°C/21°C, relative humidity of 6%0/80% and photon flux density (PFD) of 280 μmol m^{-2} s^{-1}.

To evaluate morphometric parameters, protein and free amino acids content and enzyme activities, on the 12th day of growth, the plants were treated for 48 h with aqueous solutions in which the freeze-dried potassium humate was dissolved at different concentrations expressed as carbon weight per litre: 0 (control), 1 and 5 mg C L^{-1}.

To evaluate the transcript accumulation, the HS were applied on the 13th day at 1 mg C L^{-1} for 24 h to two different control solutions: a complete nutrient solution (see above) and a nutrient solution N-deprived, where N sources KNO_3 and $Ca(NO_3)_2$ $4H_2O$ were replaced with KCl and $CaCl_{2*}2H_2O$ at the same concentrations. The expression analyses were performed by semi-quantitative reverse transcriptase-polymerase chain reaction (RT-PCR) on cDNA obtained from leaves and root tissues.

Determination of Protein Content

For the extraction of soluble protein, frozen foliar and root tissue (0.5 g) were ground in liquid nitrogen, stirred with the extraction buffer (100 mM Tris-HCl pH 7.5, 1 mM Na_2EDTA, 5 mM DTT) and centrifuged at 14,000 g. The supernatants were mixed with 10% (*w/v*) trichloroacetic acid (TCA) and centrifuged. The pellets obtained were then re-suspended in 0.1 N NaOH. The amount of total proteins in the leaves was determined by using the Bradford method [48] with bovine serum albumin as a standard. The results were expressed as mg protein/gramme fresh tissue.

Enzyme Extraction and Assay Conditions

The enzymes were solubilized from leaves by manually crashing vegetal tissues in a mortar added with a 100 mM Hepes-NaOH solution at pH 7.5, a 5 mM $MgCl_2$ solution, and a 1 mM dithiothreitol (DTT) solution. The ratio of plant material to mixture solution was 1:3 *w/v*. The extract was filtered through two layers of muslin and clarified by centrifugation at 20,000 g for 15 min. The supernatant was used for enzymatic analysis. All steps were carefully performed at 4°C, while a Jasco V-530 UV-vis spectrophotometer (Jasco Co., Midrand, Gauteng, and South Africa) was used for measuring enzyme activity.

The nitrate reductase (NR EC 1.7.1.1) activity was assayed according to Lewis and collaborators [49] and expressed as unit per milligramme protein. One unit corresponds to the production of 1 μmol of NO_2^- per minute at 25°C.

The nitrite reductase (NiR EC 1.7.1.4) activity was determined on the basis of the drop in NO_2^-concentration in the reaction medium [50]. After incubation at 30°C for 30 min, the NO_2^- content was determined colorimetrically at A*540* and the activity was expressed as unit per milligramme protein. One unit corresponds to the consumption of 1 μmol of NO_2^- per minute at 25°C.

To evaluate the glutamine synthetase (GS EC 6.3.1.2) activity, the mixture for the assay contained 90 mM imidazole-HCl (pH 7.0), 60 mM hydroxylamine (neutralized), 20 mM Na_2KAsO_4, 3 mM $MnCl_2$, 0.4 mM ADP, 120 mM glutamine and the appropriate amount of enzyme extract. The assay was performed in a final volume of 750 μL.

The enzymatic reaction was developed for 15 min at 37°C. -Glutamyl hydroxamate (gh) was colorimetrically determined by addition of 250 μL of a mixture (1:1:1) of 10% (*w/v*) $FeCl_3.6H_2O$ in 0.2 M HCl, 24% (*w/v*) trichloroacetic acid and 50% (*w/v*) HCl. The optical density was recorded at A*540* [51]. The activity was expressed as unit per milligramme protein. One unit corresponds to the production of 1 μmol mg protein per minute at 37°C.

NADH-dependent glutamate synthase (NADH-GOGAT EC 1.4.1.14) assay contained 25 mM Hepes-NaOH (pH 7.5), 2 mM L^{-1} glutamine, 1 mM -ketoglutaric acid, 0.1 mM -NADH, 1 mM Na_2 EDTA and 100 μL of enzyme extract in a final volume of 1.1 mL. GOGAT was assayed spectrophotometrically by monitoring -NADH oxidation at A*340*. The activity was expressed as unit per milligramme protein. One unit corresponds to the oxidation of 1 μmol of -NADH per minute at 37°C.

The activity of aspartate aminotransferase (AspAT EC 2.6.1.1) was measured spectrophotometrically by monitoring -NADH oxidation at A*340* at 30°C. The assay medium (final volume 2.4 mL) contained 100 mM Tris-HCl at pH 7.8, 240 mM L-aspartate (Asp), 0.11 mM pyridoxal phosphate, 0.16 mM NADH, 0.93 kU malate dehydrogenase (MDH), 0.42 kU lactate dehydrogenase (LDH), 12 mM 2-oxoglutarate and 200 μL enzyme extract [52]. The activity was expressed as unit per milligramme protein. One unit corresponds to the oxidation of 1 μmol of -NADH per minute at 37°C.

The activity of asparagine synthetase (AS EC 6.3.5.4) was determined by incubating 10 mM glutamine (Gln), 30 mM ATP, 10 mM Asp, and 10 mM $MgCl_{2,}$ 2 mM DTT, 0.1 mM EDTA, Hepes 50 mM pH 7.75 with 400 μL of the desalted extract in a total volume of 500 μL at 30°C for 60 min. The reaction was terminated with 100 μL of sulfosalicylic acid (200 mg mL^{-1}), centrifuged and an aliquot of the supernatant, adjusted to pH 7.0 with NaOH and taken for the analysis and separation of Asp and Asn by HPLC of the ortho-phthalaldehyde (OPA) reagent derivates [53]. The activity was expressed as unit per milligramme protein. One unit corresponds to the production of 1 μmol of Asn per minute at 30°C.

For the assay of citrate synthase (CS EC 2.3.3.1), the leaves were homogenized using 100 mM Tris-HCl buffer pH 8.2, containing 5 mM β-mercaptoethanol (Sigma-Aldrich, St. Louis, MO, USA), 1 mM

Na_2EDTA and 10% glycerol. The leaves were filtered and centrifuged as reported [54]. All steps were performed at 4°C. The CS enzyme was assayed adding 50 mL of oxalacetic acid 0.17 mM, 50 mL acetyl-CoA 0.2 mM, and 10 mL extract, to 3 mL of Tris-HCl 0.1 M (pH 8.0). This activity was measured spectrophotometrically at 25°C, by monitoring the reduction of acetyl-coenzyme a (acetyl-CoA) to CoA, at wavelength A*232*. The activity was expressed as unit per milligramme protein. One unit corresponds to the reduction of 1 μmol of acetyl-CoA per minute at 30°C.

To extract $NADP^+$-isocitrate dehydrogenase ($NADP^+$-IDH EC 1.1.1.42) and MDH (EC 1.1.1.37), the leaves were homogenized using 100 mM Tris-HCl buffer pH 8.2 containing 5 mM -mercaptoethanol (Sigma), 1 mM Na_2EDTA and 10% glycerol. The extracts were filtered and centrifuged as reported [54]. All steps were performed at 4°C.

For $NADP^+$-IDH activity, 50 μL of crude extract was added to 2.85 final volume of a reaction mixture containing 88 mM imidazole buffer (pH 8.0), 3.5 mM $MgCl_2$, 0.41 mM -NADP-Na salt and 0.55 mM isocitrate-Na salt. The assay was performed at 25°C following the formation of NADP (H) at A*340* [55]. The activity was expressed as unit per milligramme protein. One unit corresponds to the production of 1 μmol of NADP (H) per minute.

For MDH activity, 3.17 mL of assay mixture contained 94.6 mM phosphate buffer (pH 6.7), 0.2 mM -NADH-Na_2 salt, 0.5 mM oxalacetic acid, and 1.67 mM $MgCl_2$. MDH activity was assayed at 25°C, following the formation of NAD^+ at A*340*[54]. The activity was expressed as unit per milligramme protein. One unit corresponds to the production of 1 μmol of NAD^+ per minute.

Free Amino Acids Determination

The free amino acids were analyzed, as by the method of Seebauer et al. [56]. Fifty milligrammes of homogenous dry powder was extracted for 1 h at room temperature with 1.5 mL of a 5% (*w*/*v*) TCA solution. The sample was clarified by centrifugation, and 1.5 mL of the supernatant was analyzed for free amino acids. The amino acid analysis was accomplished by precolumn OPA derivatization of the sample followed by reverse phase separation, using an Agilent 1100 HPLC (Agilent Technologies, Palo Alto, CA, USA) equipped with a thermo-controlled auto-sampler, fluorescence detector and an Agilent HP Chemstation for data elaboration. The chromatographic conditions were described previously [57].

RNA Extraction and cDNA Synthesis

The tissue samples were ground in liquid nitrogen. Aliquots of approximately 100 mg were used for RNA extraction using TRIzol reagent (Invitrogen, Carlsbad, CA, USA) according to the manufacturer's instructions. RNA was quantified by spectrophotometric reading, and the quality was assayed by agarose gel electrophoresis. After DNAse treatment (Promega, Milan, Italy), first-strand cDNA was synthesized from 5 μg of total RNA using 200 U of MMLV Reverse Transcriptase (Promega, Milan, Italy) and oligo(dT) as a primer in 20 μL reactions, as described by Sambrook and collaborators [58].

Semi-Quantitative RT-PCR Analysis

To determine the expression level of *NR* (NCBI accession number M27821), *AS* (NCBI accession number X82849), *MDH* (NCBI accession number T27564), *CS* (NCBI accession number W49861) and *IDH* (NCBI accession number W21690), semi-quantitative RT-PCR experiments with specific primers were performed as described earlier [59]. Genes analyzed and relative primers are listed in Table 1. The constitutively expressed 18S rRNA (NCBI; Accession Number U42796.1) gene was used as an internal control of RNA quantity. PCR products were separated by electrophoresis in a 1-2.5% agarose gel stained with ethidium bromide, and quantified through the ImageJ program (ImageJ 1.38 J, Wayne Rasband, National Institute of Health, Bethesda, MD, USA). PCR reactions were performed on cDNAs obtained from two different RNA extractions from independent experiments and repeated at least three times for each cDNA. PCR products were further purified by using the QIAquick gel extraction Kit (Qiagen, Hilden, Germany) and sequenced, according to Sanger et al. [60], using the ABIPRISM original Rhodamine Terminator kit (Applied Biosystems, Branchburg, NJ, USA). The sequence comparisons were performed by using Blastx and Blastn computer programs (NCBI, National Center for Biotechnology Information).

Table 1: Specific forward and reverse primer sequences used in semi-quantitative RT-PCR

Gene	Primer forward (5′-3′)	Primer reverse (5′-3′)	Accession number
NR (nitrate reductase)	5′-TTCATGAACACTACCGACGTCG-3′	5′-GAGCCTGTACGGATACTCGGC-3′	M27821
AS (asparagine synthetase)	5′-CATCATTGAGCTCTCGCGCAGGTTAC-3′	5′-GGGGGAAATGTTATGAAGCGTTCACAA-3′	X82849
CS (citrate synthetase)	5′-GTTTGGTCATGGAGTTCTGCGTAA-3′	5′-GGAGGTACAACTTCATACAACTTGGACAC-3′	W49861
IDH (isocitrate dehydrogenase)	5′-AAACTCGAGGCTGCTTGCGTTGAGA-3′	5′-ATAATTAGCTTGCATCGAAACTGCGG-3′	W21690
MDH (malate dehydrogenase)	5′-GCCAGATTTCTGAGAGACTTAATGTCCA-3′	5′-TCGAGGCATGAGTAAGCAAGCGTCTT-3′	T27564
18S	5′-CCATCCCTCCGTAGTTAGCTTCT-3′	5′-CCTGTCGGCCAAGGCTATATAC-3′	U42796.1

Vaccaro et al.

Vaccaro et al. Chemical and Biological Technologies in Agriculture 2015 2:5, doi: 10.1186/s40538-015-0033-5

Statistical Analysis

All sets of experiments were repeated three times, and for each set three replicates were assayed. Data were the means of three replicates, and the standard deviations did not exceed 5%. The results obtained were processed statistically by the Student-Newman-Keuls test [61].

RESULTS

Effects of HS Treatments on Plant Growth

After growing for 12 days in a complete nutrient solution, the plants were transferred in a solution containing two different concentrations of HS and treated for 48 h. For seedlings supplied with 1 mg C L^{-1} of humic solution, no significant difference from control was found for leaves, roots length and fresh weight. Conversely, treatment with a greater humic concentration (5 mg C L^{-1}) produced length of both leaves and roots 8% shorter than the control seedlings (Figure 1a, b) and fresh weight (Figure 2a, b).

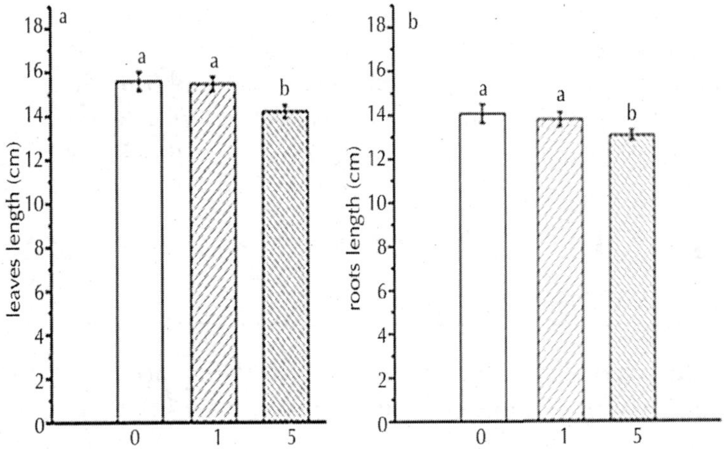

Figure 1: Length (cm) of leaves (a) and roots (b). Seedlings were grown for 14 days in a Hoagland modified complete nutrient solution (0) and treated for the last 2 days with HS concentrations of 0, 1 and 5 mg C L^{-1}. The length of roots and leaves was expressed in centimetres. Values are means of three independent determinations (±SE). Letters above bars indicates significant differences ($P<0.05$).

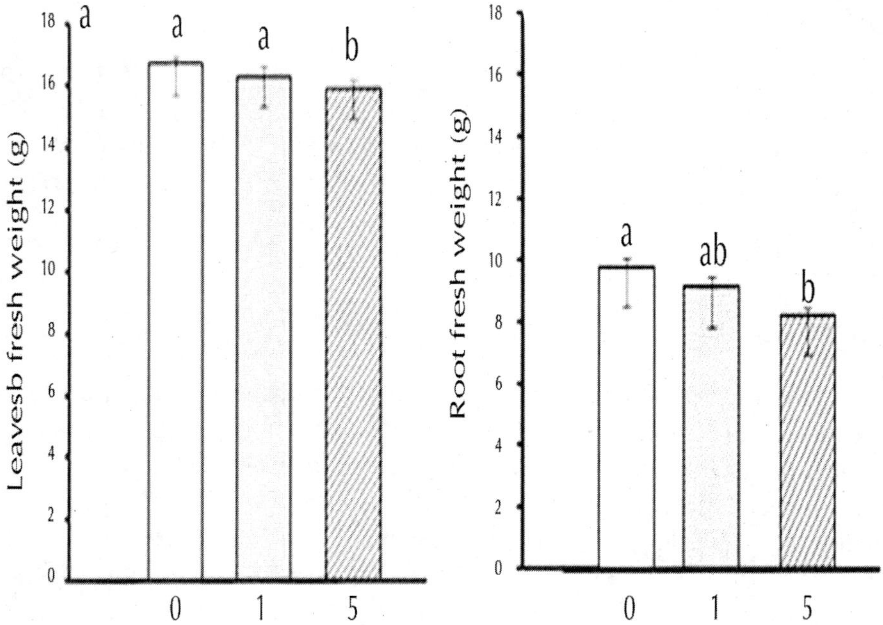

Figure 2: Fresh weight (g) of leaves (a) and roots (b). Seedlings were grown for 14 days in a Hoagland modified complete nutrient solution (0) and treated for the last 2 days with HS concentrations of 0, 1 and 5 mg C L^{-1}. The fresh weight was expressed in g. Values are means of three independent determinations (±SE). Letters above bars indicate significant differences ($P < 0.05$).

Effects of HS Treatments on Protein Content and Enzyme Activities

The amount of proteins was always significantly lower in roots than in leaves of maize seedlings. Seedlings treated with 1 mg C L^{-1} of HS showed a leaf content of protein larger than control (7%). On the other hand, a significant decrease in protein content (−21%) was observed in the leaves of seedlings supplied with 5 mg C L^{-1} HS (Figure 3a). In roots, no significant differences with respect to the control were detected at 1 mg C L^{-1} of HS and the 5 mg C L^{-1} treatment even caused a 5% decrement in protein content (Figure 3b).

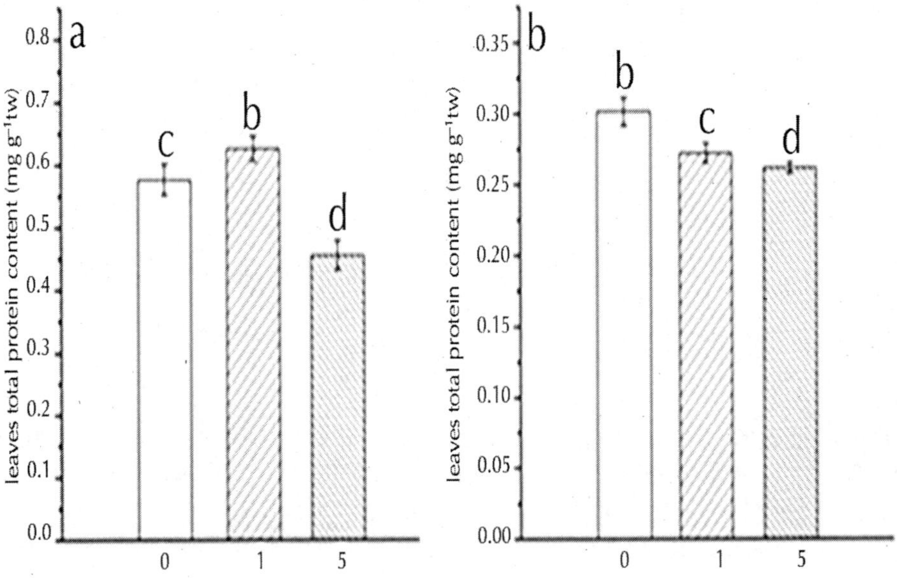

Figure 3: Total protein content in leaves (a) and roots (b).Seedlings were grown for 14 days in a Hoagland modified complete nutrient solution (0) and treated for the last 2 days with HS concentrations of 0, 1 and 5 mg C L^{-1}. The protein concentration was expressed in milligrammes of protein per gramme of fresh weight. Values are means of three independent determinations (±SE). Letters above bars indicates significant differences ($P < 0.05$).

The humic addition to the growing medium affected the activities of NR and NiR in a concentration-dependent manner (Figure 4a, b). As for NR, the application of 1 and 5 mg C L^{-1} concentrations increased the NR specific activity by about 10% and 56%, respectively (Figure 4a). A similar behaviour was found for NiR activity. In plants supplied with 1 and 5 mg C L^{-1}, the NiR activity was increased by 12% and 41%, respectively (Figure 4b).

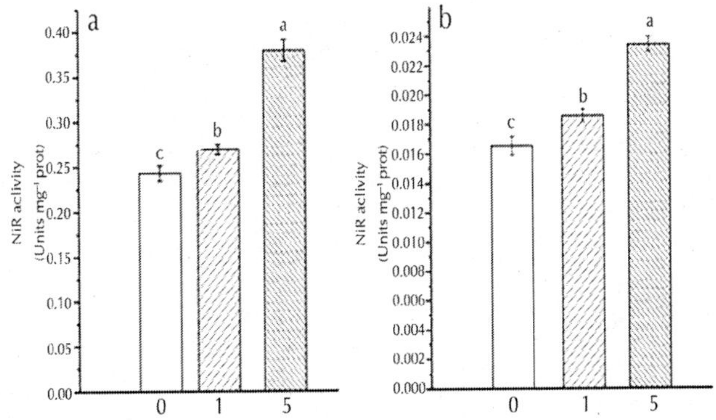

Figure 4: NR (a) and NiR (b) specific activities. Seedlings were grown for 14 days in a Hoagland modified complete nutrient solution (0) and treated for the last 2 days with HS concentrations of 0, 1 and 5 mg C L^{-1}. The specific activities were expressed in unit per milligramme protein. One unit corresponds to the production (NR) or consumption (NiR) of 1 μmol of NO_2^- per minute at the assay conditions. Values are the means of three independent determinations (±SE). Letters above bars indicates significant differences ($P<0.05$).

The GS and NADH-GOGAT enzymes showed a dose-response activity similar to those described for NR and NiR. When maize seedlings were supplied with HS at 1 and 5 mg C L^{-1}, GS and GOGAT activities increased by 11% and 30%, and by 64% and 92% respectively (Figure 5a, b).

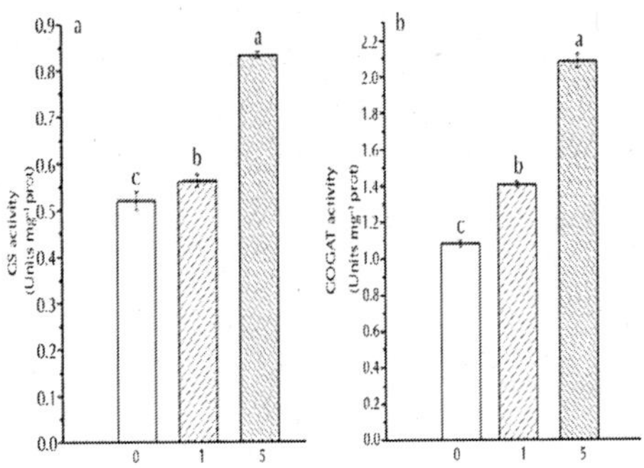

Figure 5: GS (a) and NADH-GOGAT (b) specific activities.Seedlings were grown for 14 days in a Hoagland modified complete nutrient

solution (0) and treated for the last 2 days with HS concentrations of 0, 1 and 5 mg C L^{-1}. The specific activities were expressed as unit per milligramme protein. One unit corresponds to the production of 1 μmol of γ-glutamyl hydroxamate per minute in the case of GS and to the oxidation of 1 μmol of β-NADH per minute in the case of GOGAT at the assay conditions. Values are means of three independent determinations (±SE). Letters above bars indicate significant differences ($P<0.05$).

As for AspAT, the plants supplied with 1 and 5 mg C L^{-1} of HS showed an increase in enzyme activity of 16% and 20%, respectively (Figure 6a). Conversely, the AS activity had a different trend, since 1 mg C L^{-1} caused a decrease in the AS activity (−15%) and 5 mg C L^{-1} of HS caused a significant larger stimulation (+18%) compared to the control (Figure 6b).

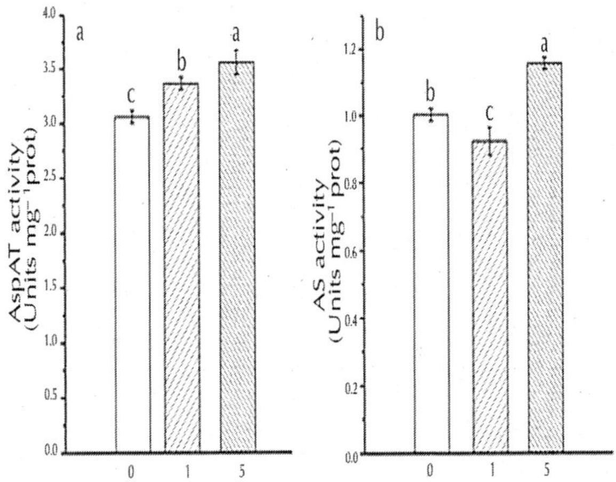

Figure 6: AspAT (a) and AS (b) specific activities. The seedlings were grown for 14 days in a Hoagland modified complete nutrient solution (0) and treated for the last 2 days with HS concentrations of 0, 1 and 5 mg C L^{-1}. The AspAT and AS specific activities were expressed in as unit per milligramme protein. For AspAT activity, 1 U corresponds to the oxidation of 1 μmol of β-NADH per minute at the assay conditions. For AS, 1 U corresponds to 1 μmol of Asn produced per minute at the assay conditions. Values are means of three independent determinations (±SE). Letters above bars indicate significant differences ($P<0.05$).

We evaluated the effects of HS on CS, NADP+-IDH and MDH of maize seedlings because of their important role in maintaining the balance between C and N metabolism. When plants were supplied with HS at 1 and 5 mg C L^{-1}, an increase of CS activity by 56% and 98%, respectively, was observed in comparison to the control (Figure 7a). A similar trend was also observed for NADP+-IDH and MDH activities (Figure 7b, c). The seedlings treated with 1 and 5 mg C L^{-1} showed an increase of both MDH and NADP+-IDH activities by 14% and 20%, and 93% and 81%, respectively.

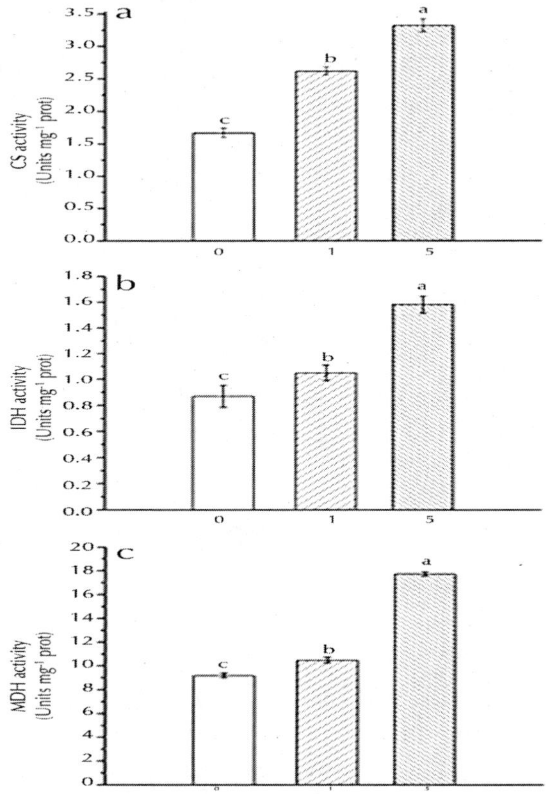

Figure 7: CS (a), NADP+-IDH (b) and MDH (c) specific activities. The seedlings were grown for 14 days in a Hoagland modified complete nutrient solution (0) and treated for the last 2 days with HS concentrations of 0, 1 and 5 mg C L^{-1}. The specific activities were expressed unit per milligramme protein. For CS, 1 U corresponds to the reduction of 1 μmol Ac-CoA per minute; for NADP+-IDH, 1 U corresponds

to 1 μmol of NADPH reduced per minute and for MDH to 1 μmol of NAD^+formed per minute at assay conditions. Values are means of three independent determinations (±SE). Letters above bars indicate significant differences ($P < 0.05$).

Effect of HS Treatment on the Amino Acid Content

The free amino acid content in leaves was measured to follow HS effects on nitrogen assimilation. Aspartate (Asp), threonine (Thr), isoleucine (Ile) and lysine (Lys) increased by 10% to 13% with 1 mg C L^{-1} HS, while the 5 mg C L^{-1} did not change the amino acids' content (Table 2). However, asparagine (Asn) behaved differently, as it was increased (41%) only with the addition of 5 mg C L^{-1} HS, while 1 mg C L^{-1} HS decreased (−11%) this amino acid in leaves as compared to control.

Table 2: Amino acid contents (μmol g^{-1}fw) in leaves

Amino acid	Treatment		
	C	1	5
Aspartate	1.74 ± 0.3	1.97 ± 0.52	1.8 ± 0.45
Threonine	2.78 ± 0.92	3.06 ± 0.96	2.76 ± 0.92
Isoleucine	1.31 ± 0.27	1.46 ± 0.35	1.3 ± 0.62
Lysine	1.36 ± 0.32	1.53 ± 0.47	1.3 ± 0.31
Asparagine	1.7 ± 0.4	1.55 ± 0.18	2.4 ± 0.39
Glutamate	2.85 ± 0.90	3.19 ± 0.96	2.71 ± 0.52

Seedlings were grown for 14 days in a Hoagland modified complete nutrient solution (0) and treated for the last 2 days with HS concentrations of 0, 1 and 5 mg C L−1. Mean values ± SD were obtained from five independent measurements.

Vaccaro et al.

Vaccaro et al. Chemical and Biological Technologies in Agriculture 2015 2:5, doi: 10.1186/s40538-015-0033-5

Effect of Humic Sample on Gene Expression

To gain further insight on the mechanisms of regulation of the enzyme previously described by HS, the expression of five genes encoding the key enzymes of N assimilation and organic acids synthesis was evaluated in leaves of both nitrate-supplied and nitrate-depleted maize seedlings, in response to 1 mg C L^{-1} HS for 24 h. The transcript accumulation of the gene encoding a NADH-NR[59] showed a significant increase after the treatment with 1 mg C L^{-1} HS in leaves of N-supplied plants, with respect to that detected in leaves of +N control plants (Figure 8). Conversely, no NR transcript accumulation was observed in the leaves of N-deprived plants of either HS-treated or control seedlings. Similarly, humic treatment induced a substantial increase of mRNA abundance in leaves of N-supplied seedlings, whereas no significant differences were detected when N-depleted seedlings were supplied with HS.

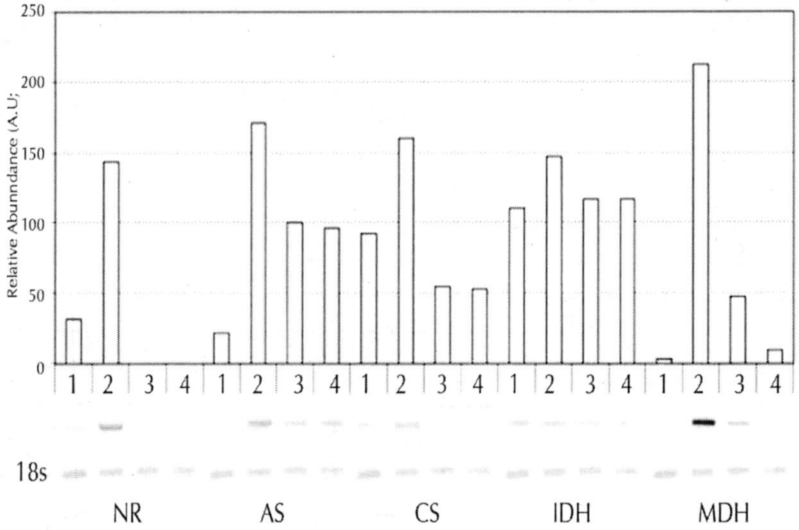

Figure 8: Analysis of the expression of NR, AS, CS, IDH and MDH in leaves. The measurements were performed by using semi-quantitative RT-PCR method in the linear range. The seedlings were grown in a control complete nutrient solution (1) and in a control N-deprived solution (3) for 14 days. One milligramme C L−1 of HS was added for 24 h after 13 days to both +N (2) and −N solutions (4). The specific transcript levels in the *y*-axis of histograms (top panel) are normalized against that of the reference gene and expressed in arbitrary units (AU).

A similar expression pattern in response to humic substances provision was also observed for the genes encoding MDH, CS and IDH enzymes which are involved in the Krebs cycle and N organication. In the case of MDH, a slight decrease of mRNA abundance was found in N-deprived leaves after humic supply. Moreover, the transcript level of these three genes was always lower in N deficiency conditions in respect to that for N-supplied plants.

DISCUSSION

Since the 1980s, several studies have highlighted the existence of biological effects of HS on plant physiology and metabolism [1], [3], [6], [25], [32]. However, only more recently, the attention of researchers has been focused on HS fractions characterized by an apparently low-molecular-weight (LMW), which seems to influence plant growth and physiology more than other fractions [3], [63]. In fact, several works focused on effects of LMW fraction on nitrogen uptake [33],[34],[37],[64] and assimilation [65], but the results obtained were often contradictory [66],[67]. Their positive effects have mainly been attributed to a large content of carbohydratic and carboxyl-C groups [32], [68].

More recently, Nardi and coworkers [36] evaluated the effect on the respiratory metabolism of maize seedlings provided by different size fractions of a soil humic acid, after their separation by preparative high-performance size-exclusion chromatography (HPSEC). The evaluation of the biological activity of size fractions showed that the size fraction with the smallest molecular size was more bioactive, followed by, in the order of, both the original bulk humic acid and the larger size fractions. Such effect was attributed to the larger flexibility of the smallest size fraction conferred by its greater content of hydrophilic components [36]. In fact, both the conformational flexibility and the hydrophilicity of this small-sized humic fraction should facilitate the release from its supramolecular structure upon the action of organic acids exuded by roots of humic molecules with plant stimulation activity. The diffusion of such bioactive molecules in solution should be less easy from humic matrices with larger and more compact conformations [37], [38], [69]. Here, we used the same bulk humic acid that was found bioactive and second only to its separated smallest size fractions in order to evaluate its effects on nitrogen assimilation in maize at both physiological and molecular levels. This species was chosen because of its worldwide economic and agronomic importance and also because of its importance as a model plant [70].

Our results indicate that the HS treatment influenced the activities of all these enzymes in a dose-dependent manner. In particular, the increase of enzyme activity in relation to the dose of HS supplied was exponential as a significant stimulation was observed when plants were treated with 5 mg C L^{-1}. A different trend was observed for AS, because its specific activity was strongly stimulated only in response to the 5 mg C L^{-1} humic solution, while it was slightly inhibited by the 1 mg C L^{-1} treatment.

The synthesis of Asn is in competition with that of Asp, Lys, Thr and Ile, which represent, together with the amino acids of the aspartate family [71], the main substrates for the synthesis of structural proteins and enzymes required for the optimal plant development [72]. However, Asn may also represent a transitory store of N to be later used to meet specific demands during plant growth [73]. Our results showed a significantly higher accumulation of Lys, Ile and Thr in leaves of seedlings grown with 1 mg C L^{-1} of HS in comparison to those observed in control plants that is in accordance with the higher content of soluble proteins measured in HS-treated leaves.

To deepen the understanding of molecular events underlying the HS effects on plant physiology, the expression of five genes encoding the enzymes involved in few key steps of N assimilation was evaluated. To discriminate N-dependent from N-independent molecular effects of HS, the transcript amount was measured in leaves of both nitrate-supplied and nitrate-depleted HS-treated seedlings. In this case, the choice of a 24-h treatment was aimed at evaluating the earlier molecular effects of humic matter. In fact, Trevisan et al. [40] showed that the gene transcription in response to HS is regulated already after a few hours of treatment.

Our findings show an increase of mRNA abundance of all genes therein studied upon HS treatment only when seedlings were grown in the presence of nitrate. On the contrary, HS did not induce any increase of transcript accumulation when the nutrient solution was depleted of nitrate. These results globally suggest that the enzyme stimulation observed after the humic supply may be generally regulated at the transcriptional level. An exception seems to be represented by the expression of the AS gene that was strongly induced after 24 h of treatment with 1 mg C L^{-1}solution, whereas the AS activity measured after 48 h was lower than for the control. This may be due either to a more rapid down-regulation of the AS activity in response to the duration of HS treatment or otherwise to the occurrence of some post-transcriptional regulatory mechanisms. Furthermore, since most of the genes analyzed are nitrate-responsive, it may be hypothesized that the HS regulation of the transcription of genes involved in nitrogen metabolism may be mediated by a nitrate-dependent signal, since HS' stimulatory effect is evident only when the nutrient solution contained nitrate.

CONCLUSIONS

Our results indicate a positive dose-dependent effect of the humic acid used here on the activities of the main enzymes involved in the reduction and assimilation of inorganic nitrogen. The enhanced bioactivity of HS may be explained with the tendency to increase the size of supramolecular aggregation of humic molecules with concentration [74] and the concomitant enhanced capacity to provide specific bioactive molecules to targeted root cell membranes [6].

Among all enzymes, only the AS showed a different trend of regulation in response to HS, possibly because of a competition of its own substrate with the biosynthesis of amino acids of the Asp family [75], thus favouring the accumulation of a large amount of Asn in seedling leaves treated with the greatest HS concentration. This is not surprising considering that Asn accumulation in young leaves is an indirect result of the restriction of protein synthesis when the metabolism is subjected to stress conditions [73]. While further studies should be certainly conducted to increase our understanding on the dose-response of HS, molecular findings suggest that the biochemical response to humic treatments may principally depend on a transcriptional mechanism of regulation, as previously hypothesized for the regulation of a maize isoform of H^+-ATPase [34] and with a more holistic approach by Trevisan et al. [46]. However, the induction of a gene expression in response to HS seems to depend closely on the presence of nitrate in the growing medium because humic matter alone was not able to up-regulate the transcription of genes. These findings suggest that HS may act as an additional signal in the regulation of gene expression mediated by nitrate, by either promoting its bioavailability or interfering with the signalling pathway that leads plants to adapt their metabolism to the nutrient availability [64],[76].

AUTHORS' CONTRIBUTIONS

SV carried out the molecular genetic studies and participated in drafting the manuscript. AE participated in drafting the manuscript and made the corrections. A participated in drafting the manuscript. AM, AP and SN conceived of the study, participated in its design and coordination and helped draft the manuscript. SQ carried out the molecular genetic studies and participated in drafting the manuscript. All authors read and approved the final manuscript.

ACKNOWLEDGEMENTS

This work was partially supported by the project COFIN 2003 no. 2003071271 funded by the Italian Ministry of University and Research (MIUR).

REFERENCES

1. Vaughan D, Malcom RE: Influence of humic substances on growth and physiological processes. In *Soil organic matter and biological activity*. Edited by Vaughan D, Malcom RE. Martinus Nijhoff/Junk W, Dordrecht; 1985:37-76.
2. Piccolo A: *Humic substances in terrestrial ecosystems*. Elsevier, Amsterdam; 1996.
3. Nardi S, Carletti P, Pizzeghello D, Muscolo A (2009) Biological activities of humic substances. In: Senesi N, Xing B, Huang PM (ed) Biophysico-chemical processes involving natural nonliving organic matter in environmental systems. Part I. Fundamentals and impact of mineral-organic-biota interactions on the formation, transformation, turnover, and storage of natural nonliving organic matter (NOM). Wiley, Hoboken, pp 305–339.
4. Muscolo A, Sidari M, Nardi S: Humic substance: relationship

between structure and activity. Deeper information suggests univocal findings. *J Geochem Explor* 2013, 129:57-63.

5. Nardi S, Pizzeghello D, Reniero F, Rascio N: Chemical and biochemical properties of humic substances isolated from forest soils and plant growth. *Soil Sci Soc Am J* 2000, 64:639-645.
6. Canellas LP, Olivares FL: Physiological responses to humic substances as plant growth promoter. *Chem Biol Technol Agric* 2014, 1:1-11.
7. Horrigan L, Lawrence RS, Walker P: How sustainable agriculture can address the environmental and human health harms of industrial agriculture. *Environ Health Persp* 2002, 110:445-456.
8. Smith EG, Young DL: The economic and environmental revolution in semi-arid cropping in North America. *Annal Arid Zone* 2000, 39:347-361.
9. Hirel B, Le Gouis J, Ney B, Gallais A: The challenge of improving nitrogen use efficiency in crop plants: towards a more central role for genetic variability and quantitative genetics within integrated approaches. *J Exp Bot* 2007, 58:2369-2387. |
10. Stevenson FJ: *Humus chemistry: genesis, composition, reactions*. 2nd edition. Wiley, New York; 1994.
11. Piccolo A: The supramolecular structure of humic substances. A novel understanding of humus chemistry and implications in soil science. *Adv Agron* 2002, 75:57-134.
12. Piccolo A, Conte P, Spaccini R, Chiarella P: Effects of some dicarboxylic acids on the association of dissolved humic substances. *Biol Fert Soils* 2003, 37:255-259.
13. Piccolo A, Conte P, Cozzolino A: Chromatographic and spectrophotometric properties of dissolved humic substances compared with macromolecular polymers. *Soil Sc* 2001, 166:174-185.
14. Orsi M: Molecular dynamics simulation of humic substances. *Chem Biol Technol Agric* 2014, 1:1-14.
15. Beiraghi A, Pourghazi K, Amoli-Diva M: Mixed supramolecular hemimicelles aggregates and magnetic carrier technology for solid phase extraction of ibuprofen in environmental samples prior to its HPLC-UV determination. *Chem Eng Sci* 2014, 108:103-110.
16. Nebbioso A, Piccolo A: Basis of a humeomics science:

chemical fractionation and molecular characterization of humic biosuprastructures. *Biomacromolecules* 2011, 12:1187-1199

17. Nebbioso A, Piccolo A: Advances in humeomic: enhanced structural identification of humic molecules after size fractionation of a soil humic acid. *Anal Chim Acta* 2012, 720:77-90.
18. Nebbioso A, Piccolo A, Lamshoft M, Spiteller M: Molecular characterization of an end-residue of humeomics applied to a soil humic acid. *RSC Adv* 2014, 4:23658-23665.
19. Niklewski B, Woiciechowski J: Uber den Einfluss der wasserlöslichen Humusstoffe auf die Entwicklung einiger Kulturpflanzen. *Z. Pflanzenernaehr Dueng. Bodenkd* 1937, 49:294-327.
20. atsky J: Uber den Einfluss der Oxy humolithen auf die mineralische Ernährung der Pflanzen. *Folia Biol* 1958, 4:443-448.
21. Khristeva LA, Gallushko AM, Gorovaya AI, Kolbassin AA, Shortshoi LP, Tkatshenko LK, Fot LW, Luk'Yakenko NV: *The main aspects of using physiologically active substances of humus nature*. VI International Peat Congress, Minnesota; 1980.
22. Nardi S, Concheri G, Dell'Agnola G: Biological activity of humic substances. In *Humic substances in terrestrial ecosystems*. Edited by Piccolo A. Elsevier, Amsterdam; 1996:361-406.
23. Vaughan D, Linehan DJ: The growth of wheat plants in humic acid solutions under axenic conditions. *Plant Soil* 1976, 44:445-449.
24. Varanini Z, Pinton R, *et al.*: Humic substances and plant nutrition. In *Progress in botany*. Edited by Behnke HD. Springer, Berlin; 1995:97-117.
25. Varanini Z, Pinton R, *et al.*: Direct versus indirect effects of soil humic substances on plant growth and nutrition. In *The rizosphere*. Edited by Pinton R. Marcel Dekker, Basel; 2001:141-158.
26. Nardi S, Concheri G, Pizzeghello D, Sturaro A, Rella R, Parvoli G: Soil organic matter mobilization by root exudates. *Chemosphere* 2000, 41:653-658.
27. Vaccaro S, Muscolo A, Pizzeghello D, Spaccini R, Piccolo A, Nardi S: Effect of a compost and its water-soluble fractions on key enzymes of nitrogen metabolism in maize seedlings. *J Agric*

Food Chem 2009, 57:11267-11276.

28. Sladky Z: The effects of extracted humus substances on growth of tomato plants. *Biolol Plant* 1959, 1:142-150.
29. Vaughan D: Effect of humic acid on the development of invertase activity in slices of beetroot tissues washed under aseptic conditions. *Humus et Planta* 1967, 4:268-271.
30. Ferretti M, Ghisi R, Nardi S, Passera C: Effect of humic substances on photosynthetic sulfate assimilation in maize seedlings. *Can J Soil Sci* 1991, 71:239-242.
31. Nardi S, Concheri G, Dellagnola G, Scrimin P: Nitrate uptake and ATPase activity in oat seedlings in the presence of 2 humic fractions. *Soil Biol Biochem* 1991, 23:833-836.
32. Piccolo A, Nardi S, Concheri G: Structural characteristics of humic substances as related to nitrate uptake and growth-regulation in plant-systems. *Soil Biol Biochem* 1992, 24:373-380.
33. Canellas LP, Olivares FL, Okorokova-Facanha AL, Facanha AR: Humic acids isolated from earthworm compost enhance root elongation, lateral root emergence, and plasma membrane H+-ATPase activity in maize roots. *Plant Physiol* 2002, 130:1951-1957.
34. Quaggiotti S, Ruperti B, Pizzeghello D, Francioso O, Tugnoli V, Nardi S: Effect of low molecular size humic substances on nitrate uptake and expression of genes involved in nitrate transport in maize (Zea mays L.). *J Exp Bot* 2004, 55:803-813.
35. Muscolo A, Sidari M, Attinà E, Francioso O, Tugnoli V, Nardi S: Biological activity of humic substances is related to their chemical structure. *Soil Sci Soc Am J* 2007, 71:75-85.
36. Nardi S, Muscolo A, Vaccaro S, Baiano S, Spaccini R, Piccolo A: Relationship between molecular characteristics of soil humic fractions and glycolytic pathway and krebs cycle in maize seedlings. *Soil Biol Biochem* 2007, 39:3138-3146.
37. Canellas LP, Piccolo A, Dobbss LB, Spaccini R, Olivares FL, Zandonadi DB, Façanha AR:Chemical composition and bioactivity properties of size-fractions separated from a vermicompost humic acid. *Chemosphere* 2010, 78:457-466.
38. Canellas LP, Dantas DJ, Aguiar NO, Peres LEP, Zsögön A, Olivares FL, Dobbss LB, Façanha AR, Nebbioso A, And Piccolo

A: Probing the hormonal activity of fractionated molecular humic components in tomato auxin mutants. *Ann Appl Biol* 2011, 159:202-211.

39. Pizzeghello D, Francioso O, Ertani A, Muscolo A, Nardi S: Isopentenyladenosine and cytokinin-like activity of four humic substances. *J Geochem Explor* 2012, 129:103-111.
40. Trevisan S, Pizzeghello D, Ruperti B, Francioso O, Sassi A, Palme K, Quaggiotti S, Nardi S: Humic substances induce lateral root formation and expression of the early auxin-responsive IAA19 gene and DR5 synthetic element in Arabidopsis. *Plant Biol* 2010, 12:604-614.
41. Trevisan S, Botton A, Vaccaro S, Vezzaro A, Quaggiotti S, Nardi S: Humic substances affect Arabidopsis physiology altering the expression of genes involved in primary metabolism, growth and development. *EEB* 2011, 74:45-55.
42. Zancani M, Petrussa E, Kraj áková J, Casolo V, Spaccini R, Piccolo A, Macrì F, Vianello A:Effect of humic acids on phosphate level and energetic metabolism of tobacco BY-2 suspension cell cultures. *Env Exp Bot* 2009, 65:287-295.
43. Zancani M, Bertolini A, Petrussa E, Kraj áková J, Piccolo A, Spaccini R, Vianello A: Fulvic acid affects proliferation and maturation phases in Abies cephalonica embryogenic cells. *J Plant Physiol* 2011, 168:1226-1233.
44. Muscolo A, Cutrupi S, Nardi S: IAA detection in humic substances. *Soil Biol Biochem* 1998, 30:1199-1201.
45. Russell L, Stokes AR, Macdonald H, Muscolo A, Nardi S: Stomatal responses to humic substances and auxin are sensitive to inhibitors of phospholipase A (2). *Plant Soil* 2006, 283:175-185.
46. Piccolo A, Conte P, Trivellone E, Van Lagen B, Buurman P: Reduced heterogeneity of a lignite humic acid by preparative HPSEC following interaction with an organic acid. Characterization of size-separates by PYR-GC-MS and 1H-NMR spectroscopy. *Envir Sci Technol* 2002, 36:76-84.
47. Nardi S, Sessi E, Pizzeghello D, Sturaro A, Rella R, Parvoli G: Biological activity of soil organic matter mobilized by root exudates. *Chemosphere* 2002, 46:1075-1081.

48. Bradford MM: A rapid and sensitive method for the quantitation of microgram quantities of protein utilizing the principle of protein-dye binding. *Anal Biochem* 1976, 72:248-254.
49. Lewis OAM, James DM, Hewitt EJ: Nitrogen assimilation in barley (Hordeum vulgare L. cv. Mazurka) in response to nitrate and ammonium nutrition. *Ann Bot* 1982, 49:39-49.
50. Lillo C: Diurnal variations of nitrite reductase, glutamine synthetase, glutamate synthase, alanine aminotransferase and aspartate aminotransferase in barley leaves. *Physiol Plant* 1984, 61:214-218.
51. Cánovas FM, Cantón FR, Gallardo F, García-Gutiérrez A, De Vicente A: Accumulation of glutamine synthetase during early development of maritime pine (Pinus pinaster) seedlings. *Planta* 1991, 185:372-378.
52. Rej R, Horder M: Aspartate aminotransferase (glutamate oxaloacetate transaminase). In *Methods of enzymatic analysis*. 3rd edition. Bergmeyer, H. U, Weinheim; 1983.
53. Lima JD, Sodek L: N-stress alters aspartate and asparagine levels of xylem sap in soybean. *Plant Sci* 2003, 165:649-656.
54. Bergmeyer HU: *Methods of enzymatic analysis*. Edited by Bergmeyer J, Marianne G. Weinheim, Basel; 1986.
55. Goldberg DM, Ellis G: Isocitrate. In *Methods of enzymatic analysis*. Edited by Bergmeyer HU. Academic, New York; 1986.
56. Seebauer JR, Moose SP, Fabbri BJ, Crossland LD, Below FE: Amino acid metabolism in maize earshoots. Implications for assimilate preconditioning and nitrogen signaling. *Plant Physiol* 2004, 136:4326-4334.
57. Henderson JW, Ricker RD, Bidlingmeyer BA, Woodward C: *Rapid, accurate, sensitive and reproducible analysis of amino acids*. Agilent Technologies, Palo Alto; 2000.
58. Sambrook J, Fritsch EF, Maniatis F: *Molecular cloning: a laboratory manual*. 2nd edition. Cold Spring Harbor Laboratory, Cold Spring Harbor, New York; 1989.
59. Quaggiotti S, Ruperti B, Borsa P, Destro T, Malagoli M: Expression of a putative high-affinity NO_3^- transporter and of an H+-ATPase in relation to whole plant nitrate transport physiology in two maize genotypes differently responsive to low nitrogen availability. *J*

Exp Bot 2003, 54:1023-1031. |

60. Sanger F, Nicklen S, Coulson AR: DNA sequencing with chain-terminating inhibitors. *PNAS* 1977, 74:5463-5467.
61. Sokal RR, Rohlf FJ: *Biometry*. 1st edition. Freeman & Co, San Francisco; 1969.
62. Stevenson FJ: *Humus chemistry: genesis, composition, reactions*. 2nd edition. Wiley, New York; 1994.
63. Nardi S, Pizzeghello D, Muscolo A, Vianello A: Physiological effects of humic substances on higher plants. *Soil Biol Biochem* 2002, 34:1527-1536.
64. Pinton R, Cesco S, Iacolettig G, Astolfi S, Varanini Z: Modulation of NO_3^- uptake by water-extractable humic substances: involvement of root plasma membrane H(+)ATPase. *Plant Soil* 1999, 215:155-161.
65. Sessi E, Nardi S, Gessa C: Effects of low and high molecular weight humic substances from two different soils on nitrogen assimilation pathway in maize seedlings. *Humic Subst Environ* 2000, 2:39-46.
66. Malcolm RE, Vaughan D: Humic substances and phosphatase activities in plant tissues. *Soil Biol Biochem* 1979, 11:253-259.
67. Mato MC, Olmedo MG, Mendez J: Inhibition of indoleacetic acid-oxidase by soil humic acids fractionated on sephadex. *Soil Biol Biochem* 1972, 4:469-473.
68. Nardi S, Pizzeghello D, Gessa C, Ferrarese L, Trainotti L, Casadoro G: A low molecular weight humic fraction on nitrate uptake and protein synthesis in maize seedlings. *Soil Biol Biochem* 2000, 32:415-419.
69. Piccolo A: The supramolecular structure of humic substances. *Soil Sci* 2001, 166:810-833.
70. Hirel B, Bertin P, Quillere I, Bourdoncle W, Attagnant C, Dellay C, Gouy A, Cadiou C, Retailliau S, Falque M, Gallais A: Towards a better understanding of the genetic and physiological basis for nitrogen use efficiency in maize. *Plant Physiol* 2001, 125:1258-1270.
71. Azevedo RA, Lancien M, Lea PJ: The aspartic acid metabolic pathway, an exciting and essential pathway in plants. *Amino*

acids 2006, 30:143-162.

72. Hirel B, Martin A, Terce-Laforgue T, Gonzalez-Moro MB, Estavillo JM: Physiology of maize I: a comprehensive and integrated view of nitrogen metabolism in a C4 plant. *Physiol Plant* 2005, 124:167-177.
73. Lea PJ, Sodek L, Parry MAJ, Shewry PR, Halford NG: Asparagine in plants. *Ann Appl Biol* 2007, 150:1-26.
74. Smejkalova D, Piccolo A: Aggregation and disaggregation of humic supramolecular assemblies by NMR diffusion ordered spectroscopy (DOSY-NMR). *Envir Sci Technol* 2008, 42:699-706.
75. Azevedo RA, Arruda P, Turner WL, and Lea PJ: The biosynthesis and metabolism of the aspartate derived amino acids in higher plants. *Phytochem* 1997, 46:395-419.
76. Nikolic M, Cesco S, Römheld V, Varanini Z, Pinton R: Uptake of iron complexed to water-extractable humic substances by sunflower leaves. *J Plant Nutr* 2003, 26:2243-2252.

Chapter 2

Effects of on-farm Composted Tomato Residues on Soil Biological Activity and Yields in a Tomato Cropping System

Catello Pane[1], Giuseppe Celano[2], Alessandro Piccolo[3], Domenica Villecco[1], Riccardo Spaccini[3], Assunta M Palese[2] and Massimo Zaccardelli[1]

[1]Consiglio per la Ricerca in Agricoltura e l'Analisi dell'Economia Agraria, Centro di Ricerca per l'Orticoltura, via dei Cavalleggeri 25, Pontecagnano, I-84098, SA, Italy

[2]Dipartimento Scienze dei Sistemi Colturali, Forestali e dell'Ambiente, Università degli Studi della Basilicata, viale dell'Ateneo Lucano 10, Potenza, I-85100, Italy

[3]Centro Interdipartimentale di Ricerca sulla Risonanza Magnetica Nucleare per l'Ambiente, l'Agro-Alimentare ed i Nuovi Materiali (CERMANU), Via Università 100, Portici, I-80055, NA, Italy

ABSTRACT

Background

The use of compost may relieve the factors that limit productivity in intensive agricultural systems, such as soil organic matter depletion and soil sickness. Concomitantly, the practice of on-farm composting allows the recycle of cropping green residues into new productive processes.

Results

We produced four vegetable composts by using tomato biomass residues in an on-farm composting plant. The tomato-based composts were assessed for their chemical, microbiological properties, and their effects on soils and plants were evaluated after their application within a tomato cropping system. Compost characteristics affected plant development and productivity through increased nutrient uptake and biostimulation functions. Soil biological activities, including basal respiration, fluorescein diacetate hydrolysis, -glucosidase, dehydrogenase, alkaline phosphatase, arylsulphatase, and Biolog community levels of physiological profiles, were differently affected by the on-farm tomato-based composts.

Conclusions

Changes in soil activity and community structure due to compost amendments were related to classes of biomolecules such as polysaccharides and lignin-derived compounds, as revealed by nuclear magnetic resonance (NMR) spectra of compost materials. The nutrient content and fertility potential of composts were positively related to the amount of tomato residues present in the feedstock.

BACKGROUND

On-farm composting is an efficient, cost-effective and environmentally safe biological process for the recycling of residual agricultural biomasses into new cropping production cycles [1]. It is a simple technology consisting of user-friendly small composting plants equipped with tools already available on a farm, where undegraded organic biomasses are transformed and stabilized through an aerobic biooxidation [2]. On-farm composting substantially contributes to solve the problem of disposing agricultural biomasses and vegetable feedstock and concomitantly provides the farmer with a self-supply of quality compost for the improvement of agricultural productivity.

Loss of soil quality is related to soil organic matter (SOM) depletion that is increased by continuous cropping without rotations, frequent soil tillage and large use of both inorganic chemical fertilizers and non-selective pesticides. Intensively exploited soils need an external supply of stabilized organic matter, such as compost, in order to counteract progressive SOM decline. Soil compost amendments contribute to the general soil quality recovery and improvement of plant growing conditions [3] by providing numerous ecosystem services, including replenishment of soil carbon stocks, increase of microbial activity and biodiversity and restoration of plant nutrition and natural soil suppressiveness [4].

In some developed horticultural areas of Southern Italy, significant amounts of agricultural wastes, such as cropping residues, unmarketable products and vegetable processing leftovers, are currently produced. They represent an important source of organic matter to be composted and returned to soil. Tomato (*Solanum lycopersicum*) green wastes from greenhouse systems produce about 15 t ha^{-1} $y^{-1}$of fresh plant residues and are among the most abundant biomasses suitable for transformation in compost.

López-Pérez et al. [5] proposed the direct incorporation of tomato residues into soil as a green biofumigating practice, but it failed in controlling nematode *Meloidogyne incognita* infestation. Risks of plant pathogen dissemination and phytotoxicity hazards are eliminated when an effective sanitation of tomato wastes is achieved through a thermophilic composting process before amendment to soils [6], [7]. Although some studies focused on tomato plant composting

[8], [7], little attention has been so far paid to assess the agronomic effectiveness of the produced compost. By assuming that on-farm composting of tomato plant wastes is the best sustainable practice to improve soil quality, our aim was to investigate (i) the effects of field compost amendments on tomato yields and resulting soil biological characteristics, (ii) the quality of on-farm composts from tomato plant residues in comparison with a commercial organic waste compost, and (iii) the molecular biomarkers which could differentiate tomato-based composts according to different amounts of tomato and other composted additives.

METHODS

On-Farm Composting

Tomato plant residues were used as main compost feedstock, while escarole (*Cichorium endivia*) residues, wood chips and mature compost as starter were also added. The four composting piles had the following compositions: C_1, 17.5% tomato, 15.5% escarole residues, 65% woodchips and 2% mature compost as starter; C_2, 25% tomato residues, 13% escarole residues, 60% woodchips and 2% mature compost as starter; C_3, 37% tomato residues, 11% escarole residues, 50% woodchips and 2% mature compost as starter; and C_4, 50% tomato residues, 48% woodchips and 2% mature compost as starter. All four raw piles were set up with an initial C/N ratio of about 30 in order to hasten the composting switch-on. The mature compost starter was a 2-year-old C_{OW}, purchased at Gesenu (Perugia, Italy). The on-farm composting process was carried out in four parallel static piles of about 6 m^3 in volume, under forced aeration, through an overall 90-day cycle that included a thermophilic and a mesophilic phase, followed by a final curing period. The on-farm composting system was assembled by using currently available tools in common farms. Mechanical aeration was provided by air injection through a net of tubes connected to a blower (0.75 KW) that was periodically activated (5 min every 3 h) with an electronic timer. Pile wetting was achieved through a PVC irrigation system, manually activated on demand (when RH $< 50\%$).

Composting temperatures were measured by thermo-sensors placed in the pile core at 15 cm from the pile bottom.

RESULTS

On-Farm Compost Characteristics

Chemical features of feedstock and composts are reported in Tables 1 and 2. Compost samples exhibited a sub-alkaline pH value (>8.0). The levels of electrical conductivity and macronutrients, including N, P and K, increased with the amount of tomato residues used, while, in all cases, the heavy metal contents detected were below risk levels according to Italian laws.

Table 1: Chemical determinations on plant residues used as composting feedstock

Residues	N (%)	P (%)	K (%)	Ca (%)	**Mg (%)**	**Na (%)**	**Mn (ppm)**	**Cd (ppm)**	**Cr (ppm)**	**Cu (ppm)**	**Pb (ppm)**	**Zn (ppm)**
Tomato	2.1	0.0062	3.04	1.46	0.20	0.57	140.94	0.253	32.33	55.81	1.06	nd
Escarole	3.8	0.0154	3.09	1.01	0.17	0.33	35.71	0.166	4.31	112.75	nd	nd
Woodchip	1.0	0.0004	0.06	0.39	0.06	0.12	7.96	0.143	6.71	5.42	nd	nd

nd, not detected. Pane *et al. Chemical and Biological Technologies in Agriculture* 2015 2:4, doi: 10.1186/s40538-014-0026-9

Table 2: Main chemical quality characteristics of the composts

Composts	pH	EC (mS cm^{-1})	N (%)	P (%)	K (%)	Ca (%)	Mg (%)	Na (%)	Mn (ppm)	Cd (ppm)	Cr (ppm)	Cu (ppm)	**Pb (ppm)**	**Zn (ppm)**
C_1	8.40	2.69	1.25	0.023	1.46	3.87	1.02	0.20	328.99	0.47	34.80	40.16	4.17	64.62
C_2	8.19	4.12	1.23	0.020	1.21	5.32	1.25	0.15	297.42	0.51	34.84	55.68	4.58	108.50
C_3	8.12	5.09	1.41	0.045	1.99	4.43	1.18	0.22	415.18	0.45	57.96	52.96	4.87	140.40
C_4	8.31	8.92	1.52	0.048	1.92	4.90	1.23	0.15	260.96	0.58	17.99	45.02	3.09	57.06
C_{OW}	8.93	5.07	2.72	0.029	1.18	6.58	0.40	0.27	427.30	0.30	16.05	45.98	28.15	247.40
Legal limits										<1.5	<100	<150	<140	<500

Pane *et al. Chemical and Biological Technologies in Agriculture* 2015 **2**:4, doi: 10.1186/s40538-014-0026-9

The ^{13}C cross polarization magic angle spinning nuclear magnetic resonance (^{13}C-CPMAS-NMR) spectra of compost materials were characterized by strong signals in the O-alkyl-C (61 to 110 ppm) region, revealing a molecular composition dominated by carbohydrates (Figure 1). In fact, the signals related to O-Alkyl-C components represent most of the organic carbon, accounting for 42.5% up to 56.3% of the total area of the nuclear magnetic resonance (NMR) spectra. The different resonances in the O-alkyl-C region are currently assigned to monomeric units in oligo and polysaccharide chains of plant tissue [9]. The intense signal around 72 ppm corresponds to the overlapping resonances of carbon 2, 3 and 5 in the pyranoside structure in cellulose and some hemicelluloses, whereas the signal at 106 ppm is assigned to the anomeric carbon 1 of the glucose unit in cellulose [10]. The shoulders localized around 62 to 65 and 84 to 88 ppm results from carbon 6 and 4 of monomeric units, respectively. The low-field resonances (higher chemical shift) of each pair indicate the presence of crystalline forms of cellulose, while the high-field ones (lower chemical shift) are assigned to either amorphous cellulose or hemicellulose structures [11]. The various O-alkyl regions could also include signals related to carbon in the propylic side chain of lignin molecules, whose smaller resonances around 62, 72 and 82 ppm, could be masked by the predominance of polysaccharides. Besides the signals usually assigned to cellulose, the spectra of different composts revealed two additional resonances around 98 and 101 ppm. These signals may be related to the di-O-alkyl-C of, respectively, monomeric units of simple carbohydrates [10] and those of either hemicellulose or pectic polysaccharide chains contained in cell walls of tomato plants, such as α-1,5 arabinan, β-1,4 galactan and α-1,4 galacturonan [12]. The broad peak in the Alkyl-C region (0 to 45 ppm) of the NMR spectra indicated the presence of alkyl chains (-CH_2- groups) derived mainly from various lipid compounds, plant waxes and polyesters. The signal at 56 ppm is associated with either the methoxyl substituent on the aromatic rings of guaiacyl and syringyl units in lignin structures or the C-N bonds in amino acid moieties [9]. Moreover, this O-alkyl region may also include the resonances related to ether and epoxy groups of plant biopolyesters. In the aromatic/olefinic-C region (111 to 145 ppm), the different resonances around 116 and 130 ppm are related to unsubstituted and C-substituted phenyl carbon pertaining to lignin monomers of guaiacyl and syringyl units [11] as well as to the ring components of plant polyphenols. The signals shown in the specific phenolic aromatic region (146 to 165 ppm) confirmed the presence of O-substituted ring carbon derived from different aromatic structures. In fact, the resonances included in the 148- to 155-ppm range are usually assigned to carbon 3, 4 and 5 in the aromatic ring in lignin components, carbon 3 and 5 being coupled to the corresponding methoxyl substituents. Conversely, the peaks found at 143 and 157 ppm in the NMR spectra of C_4 suggest the significant incorporation of polyphenol derivatives originating from tomato residues [10]. Finally, the broad signal at 173 ppm indicates the contribution of carbonyl groups of aliphatic acids and amino acid moieties in all the compost materials. The ^{13}C-CPMAS-NMR signals exhibited differences among composts (Table 3). The aliphatic alkyl C region (45 to 0 ppm) was most evident in commercial organic-waste compost (C_{OW}), followed by C_1 and C_4 then C_2 and C_3. The CH_3O/C-N region (60 to 46 ppm) was slightly variable among composts. The O-alkyl C region (110 to 61 ppm) was largely developed in C_3, whereas it was less noticeable in the remaining samples. Moreover, the intensity of the region associated with aromatic C (145 to 111 ppm) was large for C_4, decreased in the order passing from C_1 and C_2. Conversely, the spectral regions associated to phenolic C (165 to 146 ppm) was relevant in C_2 and limited in C_{OW}, while that for carboxyl C (195 to 166 ppm) was smaller in C_3 than for the rest of the other composts. Fluorescein diacetate hydrolytic (FDAH) activity resulted as the largest for C_4 and was followed, in the order, by C_3, C_2 and C_1, while C_{OW} showed the smallest value (Figure 2). Similarly, basal respiration (BR) was the greatest for the C_4 compost.

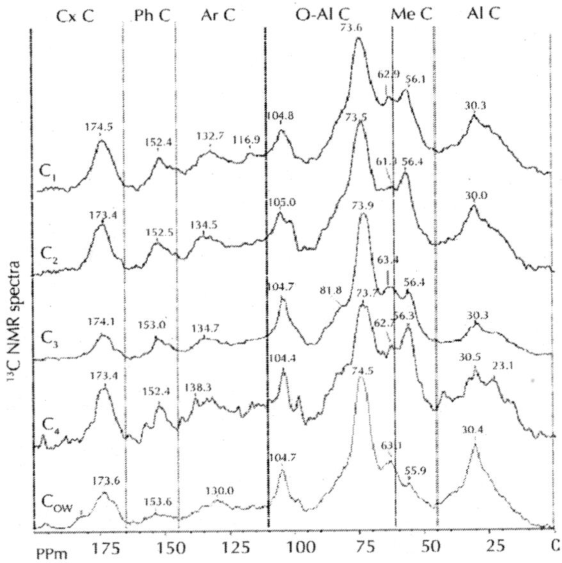

Figure 1: ^{13}C-CPMAS-NMR spectra of compost ($C_1$to C_{OW}) samples.Vertical lines delimitate six different spectral regions: aliphatic and aromatic carboxyl C (Cx C, 190 to 166 ppm); oxygen-substituted aromatic C from lignin and non-hydrolyzable tannins, phenolic and O-aryl (Ph C, 165 to 146 ppm); un-substituted and alkyl-substituted aromatic C, aryl (Ar C, 145 to 111 ppm); anomeric C and di-Oalkyl and oxidized and/or carbohydrate C, O-alkyl (O-Al C, 110 to 61 ppm); methoxyl/N-alkyl (Me C, 60 to 46 ppm); and aliphatic C, alkyl (Al C, 45 to 0 ppm).

Table 3: Relative distribution (%) of signal area over chemical shift regions (ppm) in^{13}C-CPMAS-NMR spectra of the composts

	Carboxylic-C	**Phenolic-C**	**Aromatic-C**	**O-Alkyl-C**	**CH_3O/C-N**	**Alkyl-C**
	195 to 166 ppm	**165 to 146 ppm**	**145 to 111 ppm**	**110 to 61 ppm**	**60 to 46 ppm**	**45 to 0 ppm**
C_1	8.56	4.79	13.18	43.41	11.82	8.24
C_2	7.46	4.62	13.12	44.82	12.08	17.90
C_3	5.61	4.15	11.50	52.97	11.39	14.37
C_4	6.25	3.75	14.55	42.41	13.05	19.99
C_{OW}	8.36	3.26	13.08	52.49	9.22	24.84

Pane *et al. Chemical and Biological Technologies in Agriculture* 2015 2:4, doi: 10.1186/s40538-014-0026-9

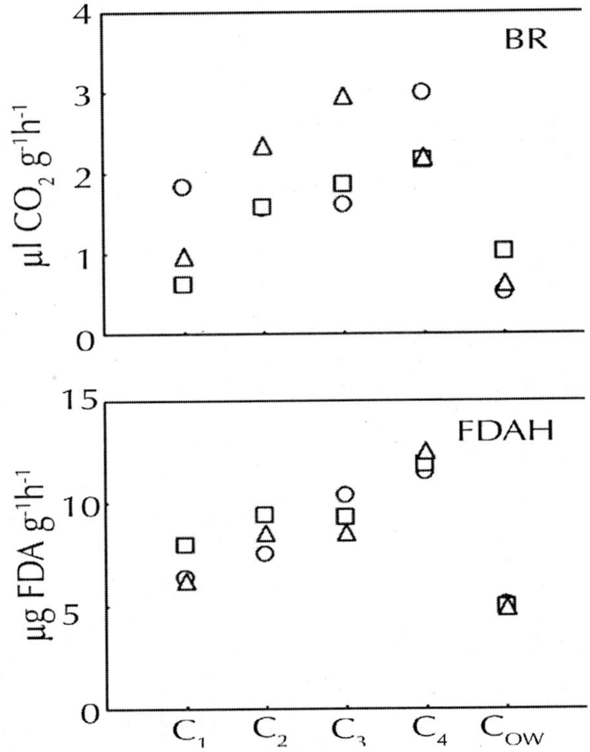

Figure 2: Biological indicators of compost stability. They were measured as basal respiration (BR) and FDA hydrolysis (FDAH) in three replicates (indicated separately with circle, triangle and square) for each compost sample.

Phytotoxicity of On-farm Composts

On-farm compost water extracts proved variable effects on cress germination index percentage (GI %) (Figure 3). In fact, while C_2 showed the lowest toxicity, the one observed for C_{OW} was the largest. Germination was increasingly repressed by on-farm compost extracts passing from C_1 to C_4. In the case of C_1, C_2 and C_3 extracts, the percent cress GI% showed a singular pattern, since it significantly increased at an intermediate concentration, while it dropped back down at the lowest concentration. $C_4$and C_{ow} extracts exhibited a dose-dependent behaviour.

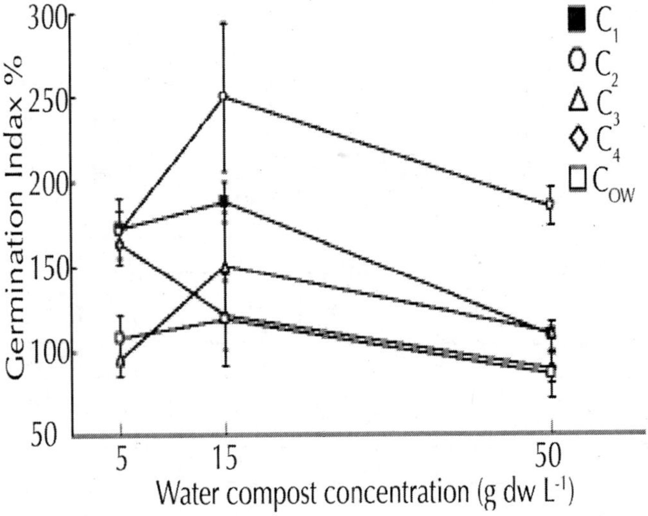

Figure 3: Evaluation of the cress germination index on water extracts of composts. C_1 to C_{OW} samples were assayed at high, medium and low concentrations (50, 16.6 and 5 g l^{-1}, respectively).

Crop Response to Soil Amendment with On-farm Composts

The commercial and total yields of C_4-treated plots were larger by about 20 t ha^{-1} than for control plots. The remaining on-farm composts (C_1, C_2 and C_3) led to increasing yields that did not differ significantly from those of untreated plots (Table 4). On-farm composts did not show any phytotoxic symptoms, whereas C_{OW} that induced the lowest yield caused a slight growth reduction in the early phases of the crop cycle. No significant differences were observed regarding the discarded production. Plant weight was significantly affected by treatments (Table 4). Similarly, berry quality (single weight, pH and optical residue) was not significantly affected by treatments, nor was plant physiologically status, which was generally observed to be at standard levels, as confirmed by chlorophyll content, likely sustained by nitrate availability throughout the tomato cycle (Figure 4). All raw compost eluates showed*in vitro* antibiosis against *Fusarium oxysporum* f. sp. *lycopersici* (data not shown).

Table 4: Effects of soil treatments on tomato cropping response

	Tomato system response						
	Green biomass (Q ha⁻¹)	**Yield**			**Quality of berries**		
		Total (t ha⁻¹)	**Marketable (t ha⁻¹)**	**Discard (t ha⁻¹)**	**Weight (g)**	**pH**	**Optical residue (°Bx)**
C_1	98.05 bc	66.6 de	62.5 c	8.51 a	75.8 a	4.15 a	3.02 a
C_2	117.08 b	83.3 bcd	75.7 bc	10.70 a	84.3 a	4.09 a	3.37 a
C_3	128.13 b	73.6 cde	69.6 bc	8.72 a	78.4 a	4.15 a	2.98 a
C_4	149.38 a	93.1 abc	88.1 ab	13.80 a	86.9 a	4.19 a	3.71 a
C_{OW}	91.81 c	60.6 e	55.1 c	8.35 a	82.0 a	4.03 a	3.03 a
M_{NR}	144.24 a	105.8 a	101.2 a	12.1 a	74.3 a	4.31 a	3.44 a
M_{SR}	148.61 a	100.6 ab	96.7 a	12.2 a	74.9 a	4.20 a	3.56 a
CTRL	109.72 bc	73.0 cde	68.2 bc	12.4 a	81.0 a	4.17 a	2.81 a

Different letters indicate significant differences (ANOVA, Duncan's test, $P \leq 0.05$).

Pane *et al. Chemical and Biological Technologies in Agriculture* 2015 2:4, doi: 10.1186/s40538-014-0026-9

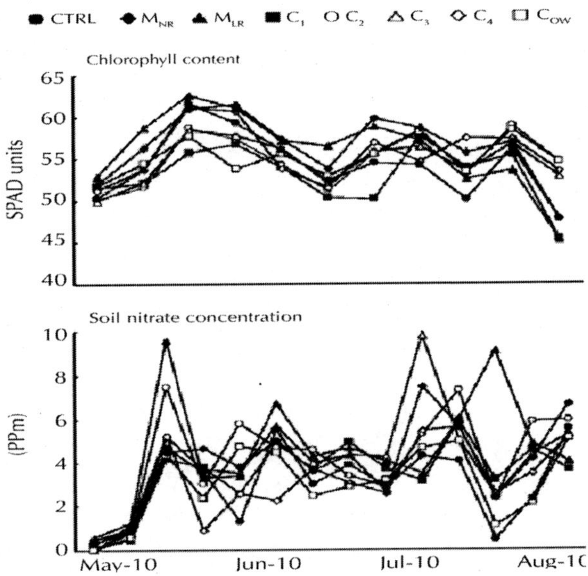

Figure 4: Plant physiological and nutritional status. It was evaluated weakly by the chlorophyll content, assessed by SPAD, and soil nitrate concentration that was available for plant nutrition.

Effects of On-farm Composts on Soil Properties

In order to assess the impact of compost amendments on soil properties, a set of biological indicators was used. The BR analyses showed an initial burst of activity due to compost amendments that approached the control over time. Levels of BR were in the following order: C_{OW}, C_4, C_1, C_2 and C_3 (Figure 5). Soil enzymatic activities that were also significantly activated by composts showed a durable effect during the whole incubation time (Figure 5). The largest values of FDAH, β-glucosidase (βGLU) and dehydrogenase (DH) activities were observed in soils amended with the municipal waste compost. Similarly, arylsulphatase (ARYL) and alkaline phosphatase (PHO) activities that resulted in large values for soils treated with tomato-based compost also showed peaks of activity only at the later stage of incubation (Figure 5). β GLU and DH, on the other hand, showed almost constant values over the whole incubation time.

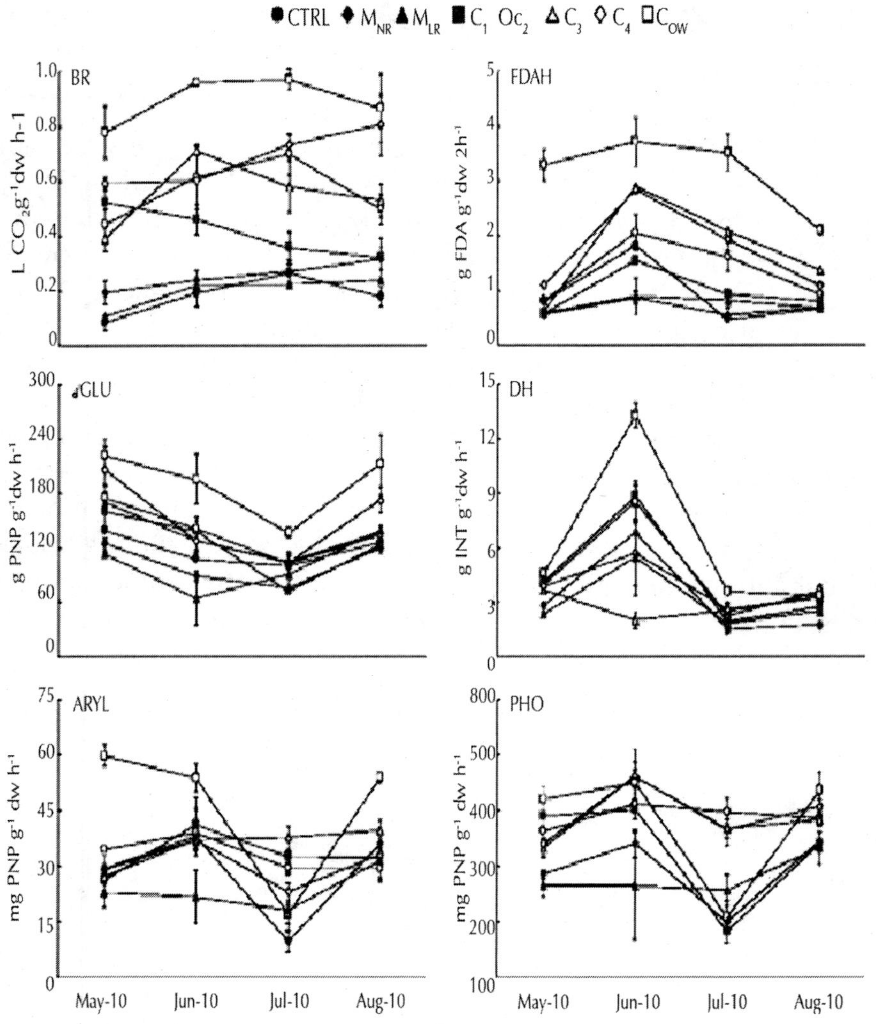

Figure 5: Changes over time in soil BR, FDAH, GLU, DH, ARYL and PHO. They were assessed monthly in experimental plots after soil treatments.

The relationships between carbon distribution in compost and soil enzymatic activities during the incubation period were elucidated by calculating the Pearson's coefficient between these two variables. In fact, coefficient profiles for cumulative regressions were generated: interestingly, it was found that polysaccharides, as well as degradation forms of lignin, produced the most significant correlations (Figure 6). Average well colour development (AWCD) and Shannon index (H') temporal shifts showed that on-farm composts significantly increased a progressive functional diversity passing from C_1 to C_4(Figure 7). The soil amended with municipal waste compost showed an intermediate behaviour as compared to that of C_2. However, the activity enhanced by compost treatments slightly, but substantially, regressed at the end of incubation time.

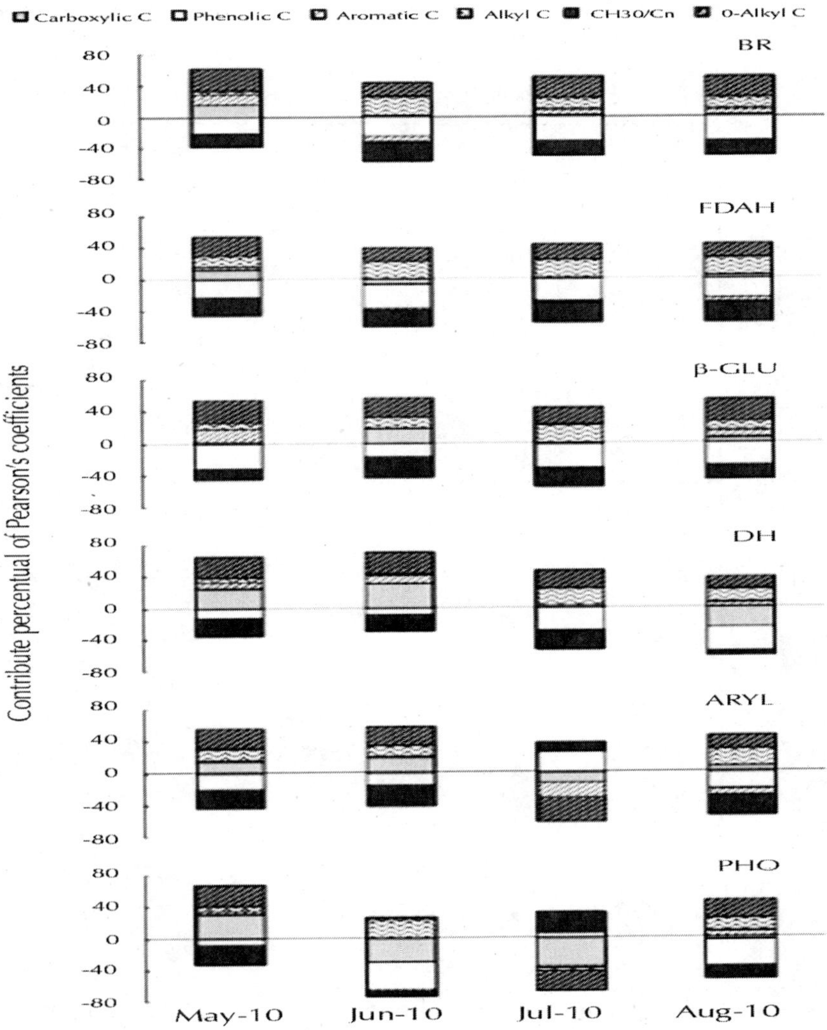

Figure 6: Cumulative regression coefficient profiles produced by cross-correlating soil activities with compost^{13}C-CPMAS-NMR regions.

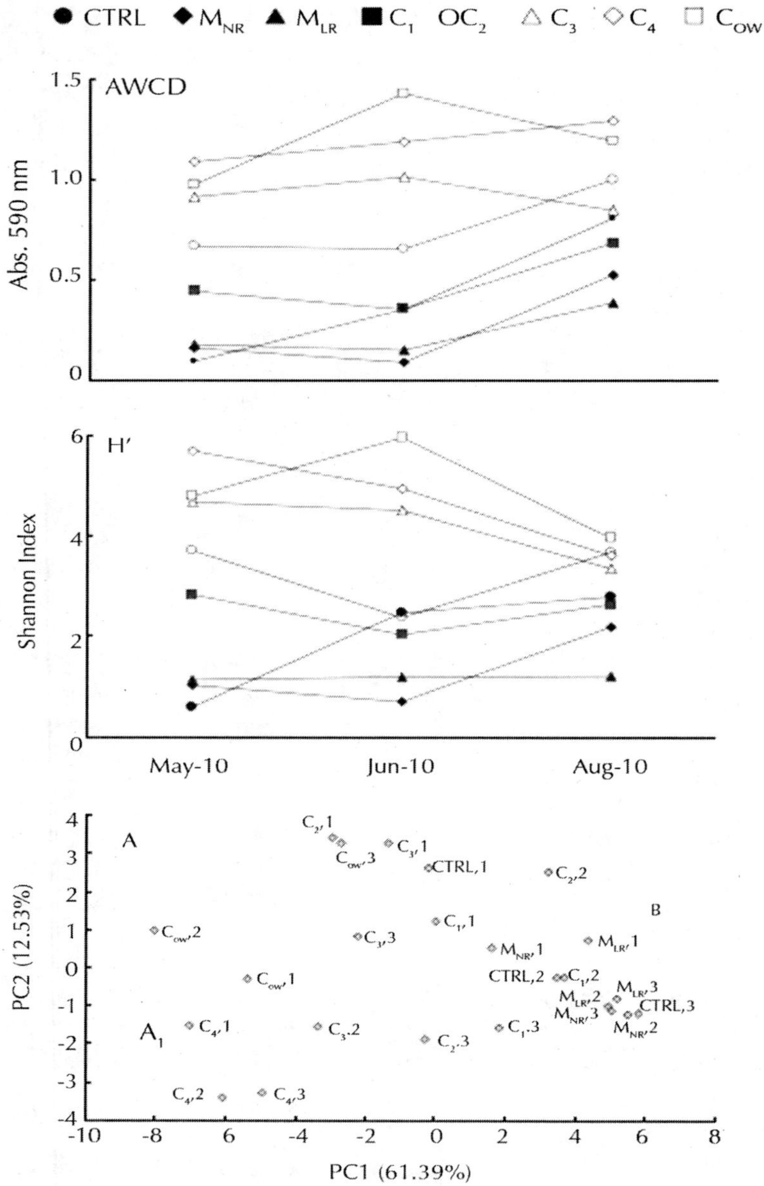

Figure 7: Changes in soil Biolog metabolic activity (AWCD), Shannon Index (H′) and biplot PCA distribution of CLPPs**.** They were assessed in experimental plots after soil treatments at three time points of incubation.

To show differences in microbial community structure, levels of carbon source catabolism were subjected to principal component analysis (PCA). The PC1 explained 61.39% of the variance, while PC2 explained only 12.53% (Figure 7). Along the PC1 axis, the compost-amended plots clustered together and resulted different from the unamended plots, though the C_4 cluster was most distant from the rest. In general, the communities in each plots grouped closely, thus indicating a little influence from sampling time. PC1-variable correlation resulted as significant ($R > 0.60$) and negative for all carbon sources (factor loadings), with the exception of hydroxy benzoic acid, d-malic acid, l-asparagine and phenylethylamine, that were not significantly correlated. Instead, only l-arginine carbon source resulted as significantly correlated ($R = 0.67$) with PC2.

DISCUSSION

Chemical Characteristics of Agricultural Composts and Stability

On-farm composting of crop residues is an effective method to produce highly humified organic matter from agricultural green wastes, while they are usefully recycled according to a concept of agricultural sustainability [13]. Moreover, these particular feedstocks significantly influenced compost properties and their ability to condition soil and plant response.

Although the ^{13}C-CPMAS-NMR spectra of composted materials indicated an overall similar C distribution, the analysis of specific signals exhibited clear differences in molecular composition. The C_1 and C_2 on-farm composts were characterized not only by cellulosic polysaccharides but also by prominent signals at 30, 56 and 152 ppm, thereby revealing significant amounts of both alkyl components and lignin derivatives. This finding suggests that the inclusion of larger initial rates of stabilizing lignocellulosic materials, represented by wood chips, promoted the incorporation of stable and recalcitrant organic components. Conversely, the NMR spectra of the C_3 compost, showed a lower content of hydrophobic alkyl and aromatic compounds and a corresponding relative increase of more biolabile O-alkyl C components. Among the on-farm composts, evidently different characteristics were found in the C_4 sample, which was made with the initial larger amount of tomato residues. Unlike the previous composts, in addition to the peak at 30 ppm, many distinct signals were shown in the broad alkyl-C region (0 to 45 ppm), thus suggesting the simultaneous presence of different alkyl chains from linear and branched fatty acids and peptidic derivatives [14]. The inconsistency between the sharp intense peaks shown at 56 ppm, as compared to the low abundance of the O-aromatic lignin components in the 148 to 155 interval, also suggested the large contribution of peptidic moieties to the global resonance in the 46 to 60 ppm region, as also indicated by the larger N content found in the C_4compost. Furthermore, the permanence of biolabile organic compounds was stressed by the peaks positioned at 43 and 98 ppm assigned, respectively, to C and C of amino acids [15] and to C1 carbon of monosaccharides components [10]. Lastly, the C distribution found in the NMR spectra of the commercial C_{OW} compost was characterized by the relative predominance of carbohydrates and alkyl-C, combined with the lowest amounts of aromatic and lignin components.

Stability is the compost property, which refers to microbial degradability of organic matter [16]. Changes in biological parameters have been indicated as reasonable and informative markers of compost stability since they shall be linkable with substrate availability for microbial growth [17]. Here, our composted residues showed increasing values of FDAH and BR according to their abundance in phenolic and aromatic-C and the amount of tomato residues used in feedstock. These hydrophobic moieties could be responsible for the maintenance of an unstable carbon reservoir formed by predominating alkyl-C and lignin-deriving compounds which was still subjected to microbial breakdown. These labile carbon pools, possibly, had already been widely removed in the commercial compost.

Effects of On-Farm Compost on Plant Growth and Productivity

Plant growth sustainability by compost refers to its quality and potential for agricultural applications [18]. Since this property, indicated as compost maturity, is closely linked to the loss of phytotoxicity [19], it can be directly assessed by evaluating the effects of compost eluates on seed germination [20]. In the current study, cress germination assay showed GI% levels exceeding the threshold value of 80%, which is indicative of mature and phytotoxin-free composts [21],[22]. In some cases, further diluted compost extracts could have stimulate seed germination. Likely, this could be due to organic molecules dissolved into compost water-extractable fractions that provide seedling development promotion [22]. However, the detection and the individuation of this kind of molecules from our on-farm composts, needs of further investigations and could be an interesting future perspective of the current research. Growth stimulation effects also may occur thank to the compost humic fraction that, for example, could be responsible for root proliferation [23], [24]. The productive response of the plants to composts may involve firstly nutrient supply and induction of well growth conditions till ripening. Here, tomato yield was observed significantly increased at levels of minerally fertilized plots, under C_4amendment, as compared to the untreated soil. C_{OW}, instead, although was the most nutrient-richest compost, leads to lowest yields because of the initial detrimental impact on the plants. These findings suggest that the nutritional value alone cannot totally explain the agronomic performances of the composts. But, other factors, such as the absence of phytotoxicity [25], the activation of soil useful microorganisms [26] or disease suppressiveness mechanisms, improvement of soil physical propriety, should be totally considered. Remarkable reduction in crop yield under plant stress due to phytotoxic composts has been widely reported [27]. Furthermore, as development of this work, the extraction and characterization of humic water-extractable fraction in composted samples could contribute to clarify these aspects.

In the current study, the antifungal activity showed by composts against *F. oxysporum* f. sp. *lycopersici*indicated, in addition, their suppressive potential, which was not possible confirm in the field since, under natural conditions, tomato wilt disease did not occurred. Anyhow, a previous study showed that some of these tomato on-farm composts (C_1 and C_4) and C_{OW} significantly suppressed soil-borne diseases caused by *Rhizoctonia solani* and *Sclerotinia minor* and suppressiveness were related to their chemical characteristics and ^{13}C-CPMAS-NMR spectra [28]. Accordingly, Yogev et al. [29] reported the ability of composted tomato plants, to suppress Fusarium wilt on melon.

Effects of On-farm Compost on Soil Biological Activities

Compost amendments represent an extraordinary food event for soil microbes that play a crucial role in the turnover of all major plant nutrients. Soil biota, in fact, is involved in the SOM cycle by activating specific enzymatic pathways, through which complex carbon structures are transformed into simple organic and inorganic molecules that can be taken up by plants. External carbon supplies, such as compost additions, cause substantial shifts in SOM chemical composition and soil biological activity profiles [30]. These changes can be followed by monitoring the evolution overtime of soil quality indicators, such as microbiological and biochemical parameters [31]. The respiration rate reflects the instantaneous microbial activation induced by labile-C compounds [30]. In this study, the BR pattern was similar to those described previously by Pane et al. [26] and Cytrin et al. [32] with an initial burst of activity induced by compost, followed by a significant decrease approaching levels of the not-amended soil. In fact, fluorescein degradation due to generalist enzymes, such as proteases, lipases and esterases [33], -glucosidase and dehydrogenase, exhibited time-shifted trends. The kinetics of these enzymes are closely related to microbial polysaccharide breakdown with the release of low molecular weight organic compounds [31] and organic matter oxidation [34]. Biolog CLPPs showed prolonged changes in microbial structures due to compost supply and were observed over incubation time, as indicated by PCA analysis. More specific enzymatic activities, such as phosphatase, which hydrolyze organic P into phosphates and arylsulphatase, which hydrolyze aromatic sulphate esters into phenols and sulphate [35] showed, instead, different behaviours. Response of soils to amendment may be affected by the molecular quality of composts. The levels of soil activity were found positively related to NMR aromatics and polysaccharides over time, suggesting the involvement of these hydrophobic moieties in the modulation of the labile carbon flux for microbial activation. Phenolic-C passes from negative to positive correlation when sulphatase and phosphatase were strongly induced in soils. A number of previous studies highlighted the role of phenolic SOM in the regulation of soil enzymatic activities [36]-[38]. Jindo et al. [39] reported the up-regulation of phosphatase and other enzymes in biochar-blended composts, concomitantly to increases in lignin polyphenol oxidation. Accordingly, Grandy et al. [37] and Leinweber et al. [38] found that strong predictive character of oxidation and the depletion rates of plant-derived lignins on the intensity of microbial transformations occurred both in less degraded systems, such as forest ecosystems, as well as in secular cropped lands. Here, on-farm composts were rich in lignin-derived compounds thanks to the large contribution of plant residues. Lignin is a relatively stable constituent of SOM that can support a long-lasting broad-spectrum soil microbial activity [40]. In the present work, solid state ^{13}C-CPMAS-NMR spectroscopy revealed differences in NMR resonance signals characteristic for alkyl C (0 to 45 ppm) and aromatic C (111 to 145 ppm) among composts that may explain the differences in induced microbiological soil activities. The alkyl C spectral region includes the aliphatic macromolecular biolipids that were reported as typical biomarkers in green waste-derived composts [41]. Instead, the aromatic C type includes peaks at 152 ppm (O-substituted C in guaiacyl and syringyl units) and 130 ppm (unsubstituted C in *p*-hydroxy phenyl rings of cinnamic units in both lignin and suberin biopolymers) and indicates that the main components of these composts are just lignocellulosic-derived molecules and hydrophobic alkyl moieties [9]. Therefore, it is possible that these lignin residues can affect indirectly biological soil properties by prolonging carbon availability to microbes over time.

Chemical Analyses and ^{13}C-cpmas-NMR Spectroscopy OF Compost Samples

Total *N* was determined according to the Kjeldahl method. The contents of P, Ca, K, Mg, Na, Cd, Cr, Cu, Mn, Pb and Zn were determined, after compost acid digestion with a microwave oven, by ICP-OES (iCAP 6000 Series, Thermo Scientific. Waltham, MA, USA). The water content of the composts was determined after drying at 105°C for 72 h. Compost water-holding capacity was determined by measuring water content held against gravity in a filter-paper-lined funnel. Electrical Conductivity (EC) and pH were determined according to the official methods of the Italian National Society of Soil Science [42].

Molecular distribution of compost organic carbon was evaluated by ^{13}C-CPMAS-NMR spectroscopy. The ^{13}C-CPMAS-NMR spectra were obtained with a Bruker AVANCE™ 300 (Bruker BioSpin GmbH,

Rheinstetten, Germany), equipped with a 4-mm wide bore MAS probe, operating at a ^{13}C resonating frequency of 75.475 MHz. Compost samples (100 to 150 mg) were packed in 4-mm zirconia rotors with Kel-F caps and spun at 13 ± 1 kHz. To account for possible inhomogeneity of the Hartmann-Hahn condition at high rotor spin rates, a 1H ramp sequence was applied in CP experiments during a contact time (CT) of 1 ms. The ^{13}C-CPMAS experiments implied a collection of 6,000 scans with 2,266 data points over an acquisition time of 25 ms and a recycle delay of 2.0 s. The Bruker Topspin 1.3 software was used to collect and process the NMR spectra. All free induction decays (FIDs) were transformed by applying a 4 k zero filling and a line broadening of 100 Hz. The areas for different ^{13}C resonances were assigned according to previous reports [4],[28],[43] into six integrating regions as follows: 0 to 45 ppm (alkyl C), 46 to 60 ppm (methoxyl C), 61 to 110 ppm (O-alkyl C), 111 to 145 ppm (aromatic C or aryl C), 146 to 165 ppm (phenolic C or O-aryl C) and 166 to 195 ppm (carboxylic C). The area of each spectral region was divided by the sum of all spectral areas in order to obtain a relative percentage.

Basal Respiration, FDA Hydrolysis and Phytotoxicity of Compost Samples

BR and FDAH were measured with a modification of method described by Pane et al. [44]. Basal respiration was from a compost (50-g dry weight) wetted with water up to 80% of its water-holding capacity and placed in a jar (500 ml) with an airtight cap. Released CO_2 was measured using a CO_2Analyser IRGA SBA-4 OEM (PP Systems, USA).

To evaluate fluorescein diacetate (FDA) hydrolysis, 2.5 g of compost was mixed with 15 ml of 0.2 M potassium phosphate buffered at pH 7.6, followed by the addition of 0.5 ml FDA solution (2 mg ml^{-1}). The mixture was shaken for 2 h in an orbital incubator and the hydrolysis reaction stopped by adding 15 ml $CHCl_3/CH_3OH$ (2:1 *v/v*). The reaction mixture was centrifuged (700×*g*) and the absorbance of the aqueous supernatant measured at 490 nm.

Composts water extracts (CWEs), prepared by vigorously shaking, were assessed for possible phytotoxicity by measuring germination and root elongation of cress (*Lepidium sativum* L. cv. Comune) under CWE treatments [45], as compared to the control H_2O. Experiments comprised three different CWE concentrations (50, 16.6 and 5 g l^{-1}) replicated 10 times. The number of seeds germinated and root length were recorded after 36 h following germination. GI% that was directly affected by phytotoxicity was then obtained by multiplying the number of germinated seeds by the relative mean root length, expressed as percentage of control accordingly to following formula [20]:

$$GI\% = \left(\frac{\text{No. of seeds germinated on CWEs}}{\text{No. of seeds germinated of water}} \right) \times \left(\frac{\text{Mean root length on CWEs}}{\text{Mean root length on water}} \right) \times 100$$

Assessment Of *In Vitro* Suppressiveness of Compost

In order to evaluate compost suppressiveness, we used the tomato fungal pathogen *F. oxysporum* f. sp.*lycopersici* Sacc. Isolated from symptomatic plants. This fungus was maintained on a potato dextrose agar medium and stored in the fungi collection of the CRA-Centro di Ricerca per l'Orticoltura (Pontecagnano, Italy). Raw, autoclaved (122°C for 22 min) and filtered (with 0.22 mm sterilized millipore membrane, following bland centrifugation to precipitate suspended cells) CWEs, further diluted in water 1:10 vol., were used to evaluate the suppressive potential of composts against *F. oxysporum* f. sp. *lycopersici*, according to the well diffusion technique as developed by El-Masry et al. [46] and slightly modified by Pane et al. [47]. Compost pathogen suppression was assessed by measuring the CWE mycelial development inhibition as a percentage of growth reduction compared to the control plates.

Field Experimental Design and Plant Parameters

Filed experiments were carried out at the experimental farm of the CRA-Centro di Ricerca per l'Orticoltura, Battipaglia (40°35′02″ N; 14°58′50″ E), Salerno, Italy, on a clay loam soil (8.8-g organic C kg^{-1}, 1.0 g Kjeldahl N kg^{-1}, pH 7.4, 34.6% sand, 36% silt, 29.4% clay, in the top 0- to 0.40-m soil layer) and the experimental design adopted was a complete randomized block with three replicates, each consisting of a plot area of 25 m^2. Eight soil treatments were compared: on-farm composts (C_1, C_2, C_3 and C_4) and municipal organic waste compost (C_{OW}) amendments, mineral normal release (M_{NR}) and mineral low release (M_{LR}) nitrogen fertilizers; untreated plots (without any fertilizers and amendment) were used as the reference control (CTRL). Composts were applied at a rate of 30 t ha^{-1}dry weight according to previous works [48], [26]. M_{NR} and M_{LR} consisted of the application of NPK synthetic fertilizers ($N = 150$ kg ha^{-1}; $P_2O_5 = 60$ kg ha^{-1}; $K_2O = 50$ kg ha^{-1}), in which nitrogen was ammonium nitrate and ENTEC®26 (a fertilizer containing 3, 4-dimethylpyrazol phosphate, a nitrification inhibitor). Composts, PK and ENTEC®26 were incorporated into the soil, 1 week before transplanting, by rotovating, at a depth of 10 to 15 cm. Tomato plantlets (cv. Stone) were transplanted (29,000 plants ha^{-1}) in double rows. During the cultivation period, plant physiological status was evaluated by assessing foliar chlorophyll contents with Minolta Chlorophyll Meter SPAD-502 (Konica Minolta Sensing INC., Japan).

At the end of the crop cycle, total and commercial production and relative percentage of discard, as well as single-berry weight, were determined on an area of about 4 m^2 for each plot.

Soil Sampling, Nitrate Determination and Microbial Activities

From May to August, after soil treatments, soil sampling was carried out by mixing 10 sub-samples that were taken from the top layer (0 to 20 cm) of each plot, sieved (2 mm), selected and stored at 4°C until biological laboratory determination. Soil moisture content was determined by measuring water content after soil drying at 50°C until constant weight. Water-holding capacity (field capacity) was determined by measuring water content held against gravity in a filter-paper-lined funnel.

Nitrate concentration in soil was analysed on weekly collected samples by colorimetric technique using Reflectoquant® strips read by a reflectometer RQflex® 10 (Merck, Darmstadt, Germany).

BR and FDAH were measured as reported above for the direct measure on composts. Soil GLU and DH activities were determined as reported by Pane et al. [44] using 4-*p*-nitrophenyl- -D-glucopyranoside

(PNP) and 2-(*p*-iodophenyl)-3-(*p*-nitrophenyl)-5-phenyltetrazolium chloride (INT) substrates, respectively. GLU was determined by adding 0.35 g of soil sample to 2 ml of 0.05 M maleate buffer, pH 5.0. The mixture was left for 5 min at 30°C, and the enzymatic reaction was started by adding 0.5 ml of 0.2 mM PNP. After incubation for 1 h at 37°C, the reaction was stopped by adding 0.5 ml of 0.5 M $CaCl_2$, 2 ml of 0.5 N NaOH and 5 ml of H_2O. After centrifugation at 1,500×*g* for 5 min at 5°C and after filtration of the aqueous phase, the absorbance of filtrates was measured at 398 nm. DH was determined by adding 0.5 g of soil sample to 1 ml of 0.2 M Tris buffer, pH 5.0. The enzymatic reaction was started by adding 0.5 ml of 0.2 mM INT. After incubation by shaking for 48 h at 37°C, the reaction was stopped by adding 10 ml of ethanol 96% *N*,*N*-dimethylformamide (1:1) and incubated in the dark by shaking for 1 h at room temperature. After centrifugation at 5,000×*g* for 5 min at 5°C and after filtration of the aqueous phase, the absorbance of filtrates was measured at 464 nm. Soil PHO and ARYL activities were determined using *p*-nitrophenyl phosphate and *p*-nitrophenyl sulphate as substrate, respectively [49]. A 0.5-g soil sample was added to 2 ml of maleate buffer, pH 6.5. The enzymatic reaction was started by adding 0.5 ml of 0.2 mM substrate. After incubation by shaking for 1 h at 37°C, the reaction was stopped by adding 0.5 ml of $CaCl_2$ and 5 ml H_2O. After centrifugation at 5,000×*g* for 5 min at 5°C and after filtration of the aqueous phase, the absorbance of filtrates was measured at 398 nm. All enzymatic analyses blanks, without addition of a reducing substrate, were also included to correct for background absorbance and the activity was determined against a calibration curve. Absorbance was measured by spectrophotometer model SpectroFlex 6600 (WTW, Oberbayern, Germany). Soil respiration and all enzymatic activities were determined on monthly sampled soils.

Biolog CLPPs were determined on soils sampled, after amendment, at beginning (May), at middle (Jun) and at end (Aug) of cropping cycle, as AWCD and H′ index, as previously developed by Pane et al. [44]. Aliquots of 100 μl of water-extracted soil sample, at a final dilution of 10^{-4} (*w/w*), were inoculated into the Eco-microplates. These were incubated at 25°C for 4 days and read, 96 h post inoculum, at 590 nm, using the Bio-Rad Microplate Reader 550 (Bio-Rad, Hercules, CA, USA).

Statistical Analysis

Tomato agronomic data were analysed by ANOVA and means were separated by Duncan's test. The relationships among biological activities detected in the amended soils and relative quantities of molecular organic ^{13}C groups of composts were assessed using a regression analysis. Biolog AWCD profiles for single substrates were computed by principal component analysis, performed on OD data of the 31 carbon sources to assess distribution biplot of all community samples.

CONCLUSIONS

This study showed the great potential of on-farm technology to produce vegetable composts with peculiar characteristics that are different from commercial composted biosolids. Nutrition and biostimulation effects may be responsible for the increased productive response to agricultural compost amendments in cropping systems. NMR profiling showed that molecular composition of on-farm composts is responsible for microbial degradability in the soil and that phenolic C could play a crucial role in modulating soil biological activities.

AUTHORS' CONTRIBUTIONS

CP, GC and DV participated in composting activities and drafted the manuscript. CP participated to the field trial. DV carried out the microbiological studies. AP and RS carried out the NMR studies and drafted the manuscript. AMP carried out chemical analysis on composts and drafted the manuscript. MZ participated in design and coordination of the study and drafted the manuscript. All authors read and approved the final manuscript.

REFERENCES

1. Maniadakis K, Lasaridi K, Manios Y, Kyriacou M, Manios T: Integrated waste management through producers and consumers

education: composting of vegetable crop residues for reuse in cultivation. *J Environ Sci Health B* 2004, 39:169-183.

2. Christian AH, Evanylo GK, Pease JW (2009) On-Farm Composting - A Guide to Principles, Planning & Operations. VCE Publ. No. 452–232.
3. Celano G, Alluvione F, Abdel Aziz M, Spaccini R (2012) the carbon dynamics in the experimental plots. Use of 13C- and 15N-labelled compounds for the soil-plant balance in carbon sequestration. In: Piccolo A (ed) Carbon sequestration in agricultural soils. A multidisciplinary approach to innovative methods. Springer, Düsseldorf, pp 107–144
4. Pane C, Spaccini R, Piccolo A, Scala F, Bonanomi G: Compost amendments enhance peat suppressiveness to *Pythium ultimum*, *Rhizoctonia solani* and *Sclerotinia minor*. *Biol Cont* 2011, 56:115-124.
5. López-Pérez JA, Roubtsova T, Ploeg A: Effect of three plant residues and chicken manure used as biofumigants at three temperatures on *Meloidogyne incognita* infestation of tomato in greenhouse experiments. *J Nematol* 2005, 37:489-494.
6. Avgelis AD, Manios VI: Elimination of tomato mosaic virus by composting tomato residues. *Neth J Plant Pathol* 1989, 95:167-170.
7. Ghaly AE, Alkoaik F, Snow A: Inactivation of *Botrytis cinerea* during thermophilic composting of greenhouse tomato plant residues. *Appl Biochem Biotechnol* 2006, 133:59-75
8. Alkoaik F, Ghaly AE: Influence of dairy manure addition on the biological and thermal kinetics of composting of greenhouse tomato plant residues. *Waste Manag* 2006, 26:902-913.
9. Spaccini R, Mazzei P, Squartini A, Giannattasio M, Piccolo A: Molecular properties of a fermented manure preparation used as field spray in biodynamic agriculture. *Environ Sci Poll Res* 2012, 19:4214-4225.
10. De Marco A, Spaccini R, Vittozzi P, Esposito F, Berg B, Virzo De Santo A: Decomposition of black locust and black pine leaf litter in two coeval forest stands on Mount Vesuvius and dynamics of organic components assessed through proximate analysis and NMR spectroscopy. *Soil Biol Biochem* 2012, 51:1-15.

11. Spaccini R, Piccolo a (2007) Molecular characterization of compost at increasing stages of maturity. 2. Thermochemolysis-GC-MS and 13C-CPMAS-NMR spectroscopy. J Agr Food Chem 55:2303–2311.

12. Rondeau-Mouro C, Crepeau M-J, Lahaye M: Application of CP-MAS and liquid-like solid-state NMR experiments for the study of the ripening-associated cell wall changes in tomato. *Int J Biol Macromol* 2003, 31:235-244.

13. Bernal-Vicente A, Ros M, Tittarelli F, Intrigliolo F, Pascual JA: Citrus compost and its water extract for cultivation of melon plants in green house nurseries. Evaluation of nutriactive and biocontrol effects. *Biores Technol* 2008, 99:8722-8728.

14. Dignac MF, Knicker H, Kogel-Knabner I: Effect of N content and soil texture on the decomposition of organic matter in forest soils as revealed by solid state CPMAS NMR spectroscopy.*Org Geochem* 2002, 33:1715-1726.

15. Kumashiro KK, Ohgo K, Niemczura WP, Onizuka AK, Asakura T (2008) Structural insights into the elastin mimetic (LGGVG)6 using solid-state 13C NMR experiments and statistical analysis of the PDB. Biopol 89:668–679

16. Iannotti DA, Pang T, Toth BL, Elwell DL, Keener HM, Hoitink HAJ: A quantitative respirometric method for monitoring compost stability. *Compost Sci Util* 1993, 1:52-65.

17. Komilis D, Kontou I, Ntougias S: A modified static respiration assay and ist relationship with an enzymatic test to assess compost stability and maturity. *Biores Technol* 2011, 102:5863-5872.

18. Som MP, Lemée L, Amblès A: Stability and maturity of a green waste and biowaste compost assessed on the basis of a molecular study using spectroscopy, thermal analysis, thermodesorption and thermochemolysis. *Biores Technol* 2009, 100:4404-4416.

19. Raj D, Antil RS: Evaluation of maturity and stability parameters of composts prepared from agro-industrial wastes. *Biores Technol* 2011, 102:2868-2873.

20. Tiquia SM, Tam NFY: Elimination of phytotoxicity during co-composting of spent pig-manure sawdust litter and pig sludge. *Biores Technol* 1998, 65:43-49.

21. Zucconi F, Pera A, and Forte M, de Bertoldi M: Evaluating toxicity of immature compost. *BioCycle* 1981, 22:54-57.

22. Tiquia SM, Tam NFY, Hodgkiss IJ: Effects of composting on phytotoxicity of spent pig-manure sawdust litter. *Environ Poll* 1996, 93:249-256.

23. Arancon NQ, Lee S, Edwards CA, Atiyeh RM: Effects of humic acids derived from cattle, food and paper-waste vermicomposts on growth of greenhouse plants. *Pedobiologia* 2003, 47:741-744.

24. Atyeh RM, Lee S, Edwards CA, Arancon NQ, and Metzger JD: The influence of humic acids derived from earthworm-processed organic wastes on plant growth. *Biores Technol* 2002, 84:7-14.

25. Levy JS, Taylor BR: Effects of pulp mill solids and three composts on early growth of tomatoes. *Biores Technol* 2003, 89:297-305.

26. Pane C, Villecco D, Zaccardelli M: Short-time response of microbial communities to waste compost amendment of an intensive cultivated soil in Southern Italy. *Comm Soil Sci Plant Anal* 2013, 44:2344-2352.

27. Woodbury PB: Trace elements in municipal solid waste composts: a review of potential detrimental effects on plants, soil biota, and water quality. *Biomass Bioenerg* 1992, 3:239-259.

28. Pane C, Piccolo A, Spaccini R, Celano G, Villecco D, Zaccardelli M: Agricultural waste-based composts exhibiting suppressivity to diseases caused by the phytopathogenic soil-borne fungi *Rhizoctonia solani* and *Sclerotinia minor*. *Appl Soil Ecol* 2013, 65:43-51.

29. Yogev A, Raviv M, Hadar Y, Cohen R, Katan J: Plant waste-based composts suppressive to diseases caused by pathogenic *Fusarium oxysporum*. *Eur J Plant Pathol* 2006, 116:267-278.

30. Tejada M, Gonzales JL, García-Martínez AM, Parrado J: Application of a green manure and green manure composted with beet vinasse on soil restoration: effects on soil properties. *Biores Technol* 2008, 99:4949-4957.

31. Serra-Wittling C, Houot S, Barriuso E: Soil enzymatic response to addition of municipal solid-waste compost. *Biol Fert Soils* 1996, 20:226-236.

32. Cytryn E, Kautsky L, Ofek M, Mandelbaum RT, Minz D: Short-term structure and functional changes in bacterial community composition following amendment with biosolids compost. *Appl Soil Ecol* 2011, 48:160-167.

33. Schnürer L, Rosswall T: Fluorescein diacetate hydrolysis as a measure of total microbial activity in soil and litter. *Appl Environ Microbiol* 1982, 43:1256-1261.

34. Fernández P, Sommer I, Cram S, Rosas I, and Gutiérrez M: The influence of water-soluble As (III) and As (V) on dehydrogenase activity in soils affected by mine tailings. *Sci Tot Environ* 2005, 348:231-243.

35. Makoi JHJR, Ndakidemi PA: Selected soil enzymes: examples of their potential roles in the ecosystem. *Afr J Biotechnol* 2008, 7:181-191.

36. Fenner N, Freeman C, Reynolds B: Observations of a seasonally shifting thermal optimum in peatland carbon-cycling processes; implications for the global carbon cycle and soil enzyme methodologies. *Soil Biol Biochem* 2005, 37:1814-1821.

37. Grandy AS, Neff JC, Weintraub MN: Carbon structure and enzyme activities in alpine and forest ecosystems. *Soil Biol Biochem* 2007, 39:2701-2711.

38. Leinweber P, Jandl G, Baum C, Eckhardt KU, Kandeler E: Stability and composition of soil organic matter control respiration and soil enzyme activities. *Soil Biol Biochem* 2008, 40:1496-1505.

39. Jindo K, Suto K, Matsumoto K, García C, Sonoki T, and Sanchez-Monedero MA: Chemical and biochemical characterisation of biochar-blended composts prepared from poultry manure. *Biores Technol* 2012, 110:396-404. Publisher Full Text

40. Tuomela M, Vikman M, Hatakka A, Itävaara M: Biodegradation of lignin in a compost environment: a review. *Biores Technol* 2000, 72:169-183.

41. Zmora-Nahum S, Hadar Y, Chen Y: Physico-chemical properties of commercial composts varying in their source materials and country of origin. *Soil Biol Biochem* 2007, 39:1263-1276.

42. Violante P (2000) Metodi di Analisi Chimica del Suolo, Angeli, F. (Eds.), pp. 536

43. Mathers NJ, Jalota RK, Dalal RC, Boyd SE (2007) 13C NMR analysis of decomposing litter and fine roots in the semi-arid Mulga Lands of southern Queensland. Soil Biol Biochem 39:993–1006

44. Pane C, Villecco D, Pentangelo A, Lahoz E, Zaccardelli M: Integration of soil solarization with *Brassica carinata* seed meals amendment in a greenhouse lettuce production system. *Acta Agric Scand Sect B Soil Plant Sci* 2012, 62:291-299.

45. Bonanomi G, Del Sorbo G, Mazzoleni S, Scala F: Autotoxicity of decaying tomato residues affects tomato susceptibility to Fusarium wilt. *J Plant Pathol* 2007, 89:219-226.

46. El-Masry MH, Khalil AI, Hassouna MS, and Ibrahim HAH: *In situ* and *in vitro* suppressive effect of agricultural composts and their water extracts on some phytopathogenic fungi. *World J Microb Biot* 2002, 18:551-558.

47. Pane C, Celano G, Villecco D, Zaccardelli M: Control of *Botrytis cinerea, Alternaria alternata* and *Pyrenochaeta lycopersici* on tomato with whey compost-tea applications. *Crop Prot* 2012, 38:80-86.

48. Zaccardelli M, Perrone D, Del Galdo A, Giordano I, Villari G, Bianco M: Multidisciplinary approach to validate compost use in vegetable crop systems in Campania Region (Italy): effect of compost fertilization on processing tomato in field cultivation. *Acta Hort* 2006, 700:285-288.

49. Stege PW, Messina GA, Bianchi G, Olsina RA, Raba J: Determination of arylsulphatase and phosphatase enzyme activities in soil using screen-printed electrodes modified with multi-walled carbon nanotubes. *Soil Biol Biochem* 2009, 41:2444-2452.

Chapter 3

Study on the Effect of Magnetic Field Treatment of Newly Isolated Paenibacillus sp.

Jie Li[1], Yanli Yi[1], Xilei Cheng[2], Dageng Zhang[1], and Muhammad Irfan[3]

[1]Key Laboratory of Preservation of Northeast Cultivated Land, Ministry of Agriculture, National Engineering Laboratory for Efficient Utilization of Soil and Fertilizer, College of Land and Enviroment, Shenyang Agricultural University, 120 dongling road, Shenyang 110866, , Liaoning, China

[2]Liaoning Academy of Environmental Sciences, Shenyang, China

[3]Bioscience and Biotechnology College, Shenyang Agricultural University, 120 dongling road, Shenyang 110866, China

ABSTRACT

Background

Symbiotic nitrogen fixation in plants occurs in roots with the help of some bacteria which help in soil nitrogen fertility management. Isolation of significant environment friendly bacteria for nitrogen fixation is very important to enhance yield in plants.

Results

In this study effect of different magnetic field intensity and treatment time was studied on the morphology, physiology and nitrogen fixing capacity of newly isolated *Paenibaccilus* sp. from brown soil. The bacterium was identified by 16S rDNA sequence having highest similarity (99%) with *Paenibacillus* sp as revealed by BLAST. Different magnetic intensities such as 100mT, 300mT and 500mT were applied with processing time of 0, 5, 10, 20 and 30 minutes. Of all these treatment 300mT with processing time of 10 minutes was found to be most suitable treatment. Results revealed that magnetic treatment improve the growth rate with shorter generation time leading to increased enzyme activities (catalase, peroxidase and superoxide dismutase) and nitrogen fixing efficiencies. High magnetic field intensity (500mT) caused ruptured cell morphology and decreased enzyme activities which lead to less nitrogen fixation.

Conclusion

It is concluded that appropriate magnetic field intensity and treatment time play a vital role in the growth of soil bacteria which increases the nitrogen fixing ability which affects the yield of plant. These results were very helpful in future breading programs to enhance the yield of soybean.

BACKGROUND

Paenibacillus genus of bacteria was first included in Bacillus genus and then reclassified to a separate genus in 1993 (Ash et al. [1993]). These bacteria found in variety of environments like soil, water, forage, rhizosphere, insect larvae, vegetable matter and in clinical samples (McSpadden Gardener [2004], (Montes et al. [2004]; Ouyang et al. [2008]; Lal & Tabacchioni [2009])). These bacteria are of prime importance in agriculture for nitrogen fixation and industrial importance due to production of antibiotics and enzymes (Mavingui & Heulin [1994]; Von der Weid et al. [2003]). These bacteria produce plant growth hormones, suppress phytopathogens and solubilize organic phosphate (Mavingui & Heulin [1994]; Lebuhn et al. [1997]; Pires & Seldin [1997]).

Nitrogen is very essential nutrient for the growth of plants. So, these bacteria fix nitrogen from the air and provide this nitrogen to plants in the form of ammonium ions or other nitrogenous compounds essential for growth. From this symbiotic association, plant provides some organic compounds synthesized from photosynthesis (Sawada et al. [2003]). These bacteria not only fix the nitrogen but also enrich the soil fertility, increase plant production, and improve the quality, degrade organic pollutants and production of vitamin B series compounds (Sierra et al. [1999]; Agus et al. [2000]). The nitrogen deficiency was recovered by these rhizobia (Fisher & Long [1992]). In this process, plant produced some reactive oxygen species including the hydrogen peroxide and hydroxyl radicals and superoxide anion by defence reaction (Lamb & Dixon [1997]; Santos et al. [2001]). So it was necessary to study the rhizobia catalase, peroxidase and superoxide dismutase active changes.

A lot of research showed that the magnetic treatments have certain stimulative effect on crop production and development and, it also affect the genetic quality of seeds ((Zhu et al. [1996]; Liu et al. [1996]; Yan et al. [1997]; He et al. [1999]; Mao et al. [2002]); Jia et al. [2002]; (Liu et al.[2003])). Enzyme as protein with catalytic activity has an important role in the life process, and as a catalyst it was increasingly being attention (Cheng et al. [2007]). Magnetic field on the influence of the enzyme activity has been reported (He et al. [1998]; Li et al. [2007]; Hua et al. [2008]), and this area now attracts more and more people's attention, but most of these studies focused on animals, plants and very little research on bacteria. So this study was aimed to check the effect of magnetic field on soybean rhizobia isolated from brown soil and their enzyme activities (peroxidase, catalase and superoxide dismutase) under the influence various intensity of magnetic treatment.

METHODS

Materials

The Brown soil samples were collected from Shenyang Agriculture University, Shenyang Liaoning P.R. China. The samples were kept in sterile plastic bags and transferred aseptically to the lab.

Isolation of *Paenibacillus*

The *Paenibacillus* sp. were isolated using standard procedures, and were purified by repeatedly streaking the bacteria on yeast extract-mannitol agar (YMA) medium (Vincent [1970]) and stored at 4°C.

Molecular Identification of *Paenibacillus*

Genomic DNA of the newly isolated bacterial strain was extracted by method as described by Ausubel et al. ([1994]). The DNA was amplified using universal primers 27 F: 5′ -GAGAGTTTGATCCTGGCTCAG-3′ and 1492R:5′ -GGYTACCTTGTTACGACTT-3′. PCR reactions were performed in 50 l volume containing 1 μL template DNA, 4 μL $MgCl_2$ (25 mmol/L), 5 μL 10× PCR buffer (Mg^{2+}free), 4 μL dNTP(10 mmol/L), 1 μL of each primer (10 μmol/L), 0.5 μL of TaqDNA polymerase (5u/μL) and 33.5 μL ddH_2O. PCR amplification conditions as follows: Initial denaturation at 94°C for 5 min followed by 30 cycles of denaturation at 94°C for 30 s, annealing at 55°C for 30 s and extension 72°C for 1 min, final extension at 72°C for 10 min. Amplification products were separated by 1 · 0% agarose gel electrophoresis and visualized under UV light after staining with ethidium bromide. The amplified 16S rRNA gene was sequenced using ABI 3730xl DNA Analyzer (Applied

Biosystems, USA). The sequences were identified based on similarity using the Basic Local Alignment Search Tool (BLAST) program National Centre for Biotechnology Information (NCBI) online standard (http://www.ncbi.nlm.nih.gov/).

Magnetic Treatment of Soil

The soil was treated by magnetic field in 100 mT, 300 mT and 500 mT for 5 min, 10 min, 20 min and 30 min respectively. Soybean was planted in the treated soil samples using phosphate and potash fertilizers (75 mg $kg^{-1}P_2O_5$; 75 mg kg^{-1} K_2O). After harvestation the plants and soil was used to determine the soybean nodulation and nitrogen fixation capacities.

Magnetic Treatment of ***Paenibacillus*** Sp.

The *Paenibacillus* sp. was inoculated in 100 mL of YMA medium, incubated at 28°C for 36 h with agitation speed 200 rpm. The cell growth was measured by taking OD at 520 nm. After the cell growth, 25 mL of *Paenibacillus* sp. cell suspension was taken in a test tube and treated it with different magnetic fields like 100, 300 and 500 mT with different time period such as 0, 5, 10, 20 and 30 min. Each experiment was conducted in triplicates and *Paenibacillus* sp. without magnetic treatment was taken as control.

Enzyme Assay

The *Paenibacillus* sp. broth was centrifuge at 5000 × g, 4°C for 10 min. After centrifugation the supernatant was discarded and the pellet was suspended in 50 mmol L^{-1} phosphate buffer (pH 7.0) and then subjected to sonication. The homogenate solution was centrifuged for 10 min at 10000 × g, 4°C. After centrifugation, the supernatant was used for determination of peroxidase (POD), superoxidase dismutase (SOD) and catalase (CAT) activities. Catalase activity was assay of hydrogen peroxide based on the formation of its stable complex with ammonium molkbdate and the OD was measured at 405 nm (Fang et al. [2004]). One unit of catalase activity was defined as the decomposition of 1 μ mol of hydrogen peroxide per minute under standard assay conditions.

Peroxidase activity was determined by hydrogen peroxide-dependent oxidation of guaiacol. Samples were mixed with guaiacol solution (20 mmol/L guaiacol in 0.1 mol/L phosphate buffer (pH 6.8) and 0.03% (v/w) hydrogen peroxide) (Bergmeger et al. [1983]). Increase in absorbance at 470 nm was recorded using UV-visible spectrophotometer. One unit of POD activity was defined as the change in absorbance of 0.01 per minute at room temperature. Total SOD activity was assayed by the inhibition of the photochemical reduction of pyrogallol (PAPG) by following the photo reduction of nitroblue tetrazolium (Cai et al. [2006]). One unit of SOD activity was defined as amount of enzyme producing a 50% suppression of PAPG reduction. All the Enzyme specific activity is expressed as U/ml.

Total Nitrogen Determination

Total plant nitrogen (N) concentration was analysed with Kjeldahl determination and colorimetric method as described by Baethgen and Alley (Baethgen & Alley [1989]). Nitrogen fixed was calculated as the total plant nitrogen content at harvest, minus the total nitrogen content at the start of the treatments.

Statistical Analysis

The data obtained after experimentation was statistically evaluated using ANOVA at significance level of $p<0.05$ by using computer based programme SPSS.

RESULTS AND DISCUSSION

Molecular Identification of *Paenibacillus* Sp.

The newly isolated bacterial strain was identified by molecular techniques using 16S rDNA sequencing. Product of 1442 bp was obtained after PCR amplification (Figure 1). The sequencing result was compared by BLAST for homology analysis. The isolated strain had 99% similarity with *Paenibacillus* sp. 9-2AIA (FN397529.1), *Paenibacillus* sp. Gi-691

(EU497639.1), *Paenibacillus chibensis strain ZYb3* (FJ432004.1), 98% similarity with *Paenibacillus sp. C-2* (KF479638.1),*Paenibacillus* sp. CC-YHH111 (JN806383.1), *Paenibacillus* sp. BM-7 (AY635866.1), 97% similarity with *Paenibacillus* sp. E18 (FJ899682.1), *Paenibacillus* sp. D27 (KF479657.1), *Paenibacillus favisporus* isolate MKI10 (EF173324.1) and 96% similarity with *Paenibacillus rhizosphaerae* strain CECAP16 (AY751755.1) and *Paenibacillus* sp. SSG-1 (KF750627.1) as shown in phylogenetic tree (Figure 2).

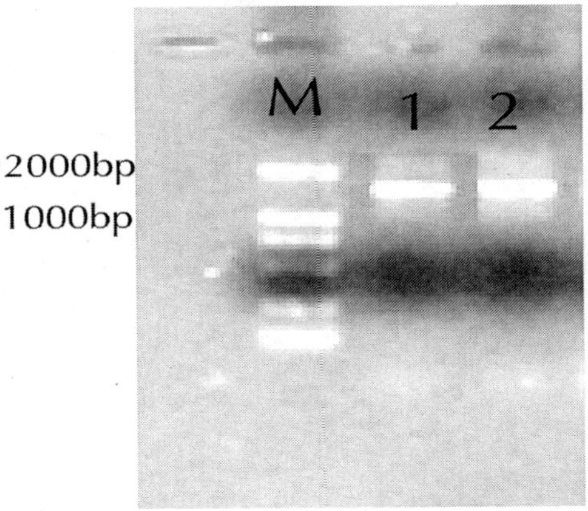

Figure 1: Amplified product of 16S rDNA from newly isolated bacterial species.

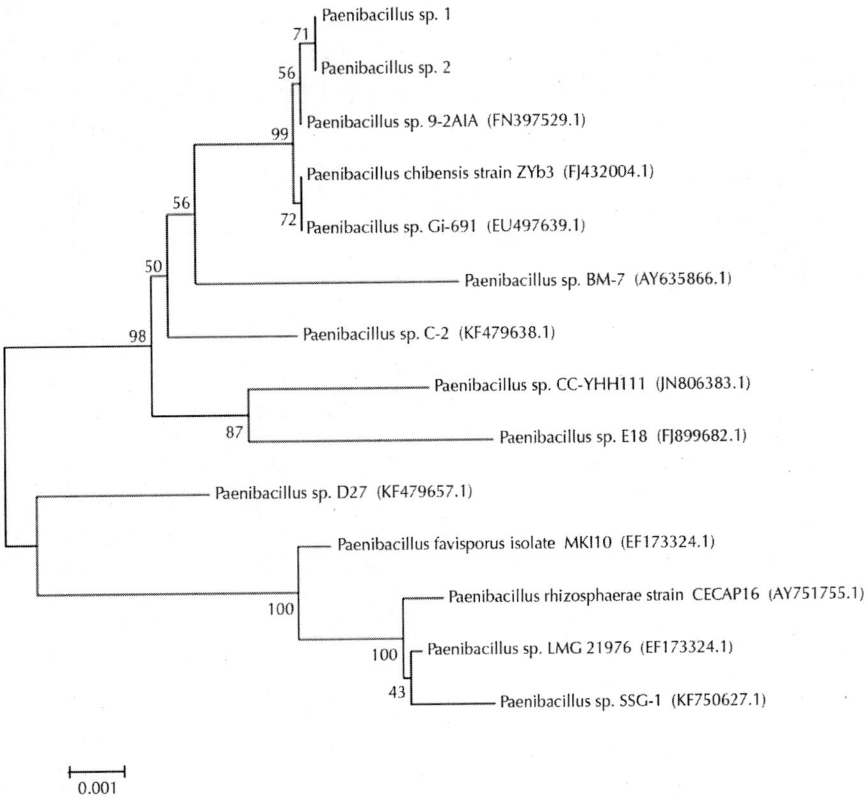

Figure 2: Neighbor-joining phylogenetic analysis of 16S rDNA sequences (1442 bp) of newly isolated*Paenibacillus*sp. Compared with the sequence of standard strains. The bar represents a 0.002% of nucleotide variance.

Effect of Magnetic Field Treated Soil Soybean Nodular and Nitrogen Fixation

The soybean was planted in the brown soil treated with magnetic field, and the plants were harvested to measure the nodular situation and nitrogen content. The soybean was planted in the absence of nitrogen fertilizer and the difference in nitrogen content before and after treatment was used to compare the treatment effects. Thus we can use plant nitrogen to respond to the difference in the amount of nitrogen-fixing performance. Results shown in the Table 1 described that the soybean plant root dry weight, effective number of *Paenibacillus* sp.

and total nitrogen content which treated by 100 mT, 300 mT magnetic field in plant were significantly higher as compared to control (CK). Results of this study was in agreement with Jing et al. (Jing et al.[1992]) who reported that appropriate intensity of magnetic field increased 4–5 time nitrogen fixing efficiency of the *Bradyrhizobium japonicum* 005. High treatment (500 mT) of magnetic field resulted decreased number of effective rhizobia thus leading to decreased total nitrogen content. This low nitrogen content leads to less growth because nitrogen is the main nutrition factor that influences the growth in plants (Marschner [1995]; Barker & Bryson [2007]). Nitrate is inorganic nitrogen which is dominant in agricultural soils and present in three or more times higher than natural soils (Hagedorn et al. [2001]; Owen & Jones [2001]).

Table 1: Effect of magnetic field treated soil on soybean nodular and nitrogen content

Treatment		Bacterial dry weight		Effective number of root nodule		Total nitrogen	
Magnetic field (mT)	Time (min)	Weight (g dry wt pl^{-1})	Percentage change (%)	Amount (No./plant)	Percentage change (%)	Content (mg/g)	Percentage change (%)
Control (CK)		0.06^{h}	-	7 ± 1.23	-	3.23 ± 0.12	-
100	5	0.20^{c}	233	24 ± 3.47	243	3.46 ± 0.13	7.12
	10	0.19^{d}	217	18 ± 1.76	157.	3.51 ± 0.12	8.67
	20	0.17^{e}	183	17 ± 1.84	143	3.35 ± 0.11	3.72
	30	0.15^{f}	150	16 ± 1.34	129	3.25 ± 0.12	0.62
300	5	0.27^{a}	350	37 ± 1.63	429	4.33 ± 0.17	34.06
	10	0.26^{b}	333	28 ± 2.95	300	4.55 ± 0.15	40.87
	20	0.15^{f}	150	24 ± 2.58	243	3.58 ± 0.13	10.84
	30	0.09^{g}	50	13 ± 1.72	86	3.51 ± 0.16	8.67
500	5	0.04^{j}	−33	7 ± 1.25	0	3.34 ± 0.13	3.41
	10	0.06^{h}	0	7 ± 1.13	0	3.25 ± 0.12	0.62
	20	0.05^{i}	−17	6 ± 1.13	−14	3.17 ± 0.11	−1.86
	30	0.05^{i}	−17	7 ± 1.12	0	3.14 ± 0.13	−2.79

The different letters show significant difference ($P < 0.05$).

Li et al.

Li et al. Botanical Studies 2015 56:2, doi: 10.1186/s40529-015-0083-9v

Effect of Magnetic Treatment on Generation Time of *Paenibacillus* Sp.

In order to check the effect of magnetic field on *Paenibacillus* sp. morphology and physiology, all the effects were studied before and after treatment using various magnetic intensities and treatment time. After the magnetic field treatment, *Paenibacillus* sp. I and *Paenibacillus* sp. II quantity and generation time were changed. On magnetic treatment, the number of *Paenibacillus*sp. Was increased as compared to control (CK). This effect was changed with respect to magnetic intensity and time duration. Results (Table 2) showed that magnetic intensity of 300 mT had strong effect on the population of *Paenibacillus* sp. The growth of *Paenibacillus* sp. was influenced by magnetic intensity and magnetic treatment time. By increasing the treatment time, number of*Paenibacillus* sp. was increased and further increase in treatment time resulted decline in*Paenibacillus* sp. growth. Of all the treatment time, 10 minutes of magnetic treatment gave better yield. The generation time of 100 mT, 300 mT magnetic field intensity treatments were shortened as compared to control (CK) while 500 mT magnetic field intensity treatments has no significant effect on the generation time as compared to control. These results indicated that magnetic field treatment significantly enhance the *Paenibacillus* sp. population in shorter time. Cheng and Yi (Cheng & Yi [2009]) reported that magnetic field intensity of 300 mT had a significant positive effect on the generation time of slow-growing rhizobium (USDA110) and fast-growing rhizobium (USDA191).

Table 2: The magnetic treatment of soybean purification number and generation of rhizobium time influence

Treatment		***Paenibacillus*sp. I**		***Paenibacillus*sp. II**	
Magnetic field (mT)	**Time (min)**	**Number**	**Generation of*Paenibacillus*sp. (h)**	**Number**	**Generation of*Paenibacillus*sp. (h)**
CK		42	10.0	40	3.9
100	5	65	8.6	75	3.0
	10	101	8.2	68	3.1
	20	79	8.4	62	3.3
	30	55	8.6	64	3.5

300	5	85	8.0	122	3.1
	10	127	8.0	83	2.8
	20	95	8.1	77	3.1
	30	66	8.1	81	3.1
500	5	94	9.6	115	3.8
	10	102	9.6	84	3.9
	20	86	9.8	82	3.9
	30	42	9.7	78	4.0

Li et al.

Li et al. Botanical Studies 2015 56:2, doi: 10.1186/s40529-015-0083-9

Effect of Magnetic Treatment on Morphology of the *Paenibacillus* Sp.

The effect of magnetic field on morphology of *Paenibacillus* sp. was also studied as shown in the Figures 3 and 4. Results showed that morphology of *Paenibacillus* sp. was affected by magnetic treatment. Only 10 minutes of magnetic treatment time with intensity of 300 mT and 500 mT was studied. In 300 mT treatment the cell becomes thin and longer while in case of 500 mT treatment the cell membrane was ruptured and the cytoplasmic sap was released leading to the death of the*Paenibacillus* sp. These results revealed that suitable processing time with specific intensity can promote the growth of the *Paenibacillus* sp. but high and longtime magnetic processing inhibited the growth thus leading to the death of *Paenibacillus* sp. Our findings were similar as reported by Fadel et al. (Fadel et al. [2003]). Various reports suggested that exposure of microorganisms to magnetic field caused changes in morphology and growth (Mohamed et al. [1997]; Gaafar et al. [2006]).

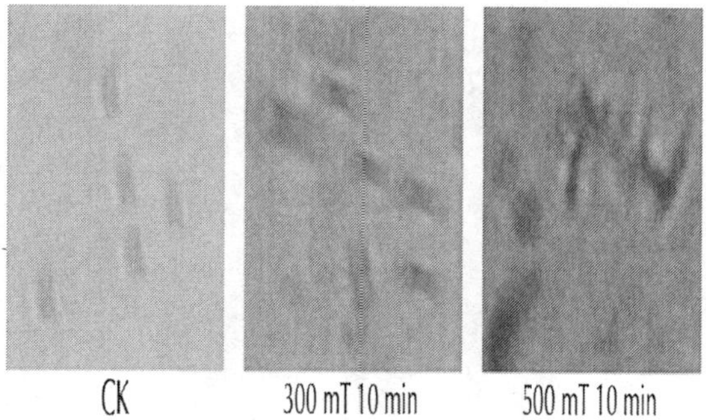

Figure 3: Effects of magnetic fields on the morphology of*Paenibacillus*sp.

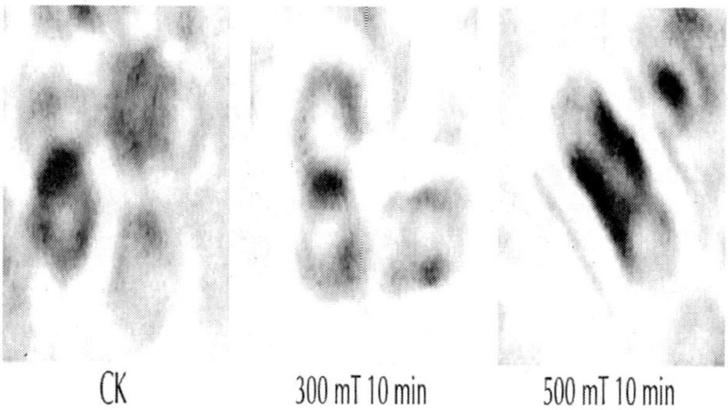

Figure 4: Effect of magnetic fields on the morphology of*Paenibacillus*sp.II.

Effect of Magnetic Field Treatment on Enzyme Activity of *Paenibacillus* Sp.

The influence of magnetic field was also checked on the catalase, peroxidase and superoxide dismutase activity of *Paenibacillus* sp. I and II. The catalases activity of *Paenibacillus* sp. I and II were enhanced by 300 mT intensity as shown in Figure 5. *Paenibacillus* sp. I showed more catalase activity as compared to *Paenibacillus* sp. II. The effective intensity and magnetic treatment time for *Paenibacillus* sp. I and II were 300 mT intensity for 10 and 20 minutes respectively. Both 100 mT and 300 mT increased the catalase activity while 500 mT decreased catalase activity as compared to control. The highest growth rate of *Paenibacillus* sp. I and II was 77% and 95% respectively. These results suggested that magnetic treatment time and intensity had strong influence on metabolic activity of *Paenibacillus* sp.

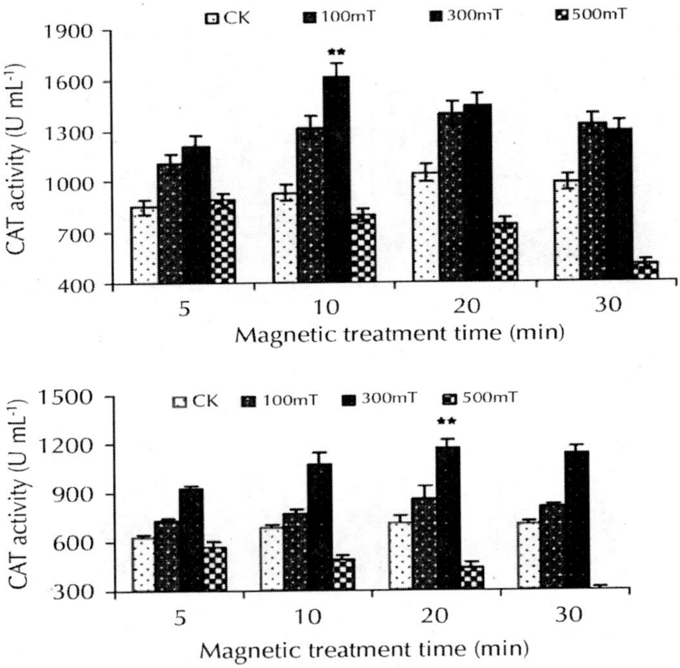

Figure 5: Effect of magnetic fields on CAT activity of*Paenibacillus*sp.I (upper) and*Paenibacillus*sp.II (lower). Error bars indicate SD among triplicates. Stars represent significance level at $P>0.05$.

Figure 6 showed the influence of different magnetic field on peroxidase activity of *Paenibacillus* sp. The 100 and 300 mT magnetic field treatment have growth trend to the peroxidase activity of*Paenibacillus* sp. The highest growth rate of *Paenibacillus* sp. I and II was 68% and 203% in 10 min and 30 min magnetic treatment respectively. Both *Paenibacillus* sp. showed maximum peroxidase activity at 10 min of treatment time with intensity of 300 mT. This activity behavior is almost similar to that of catalase activity.

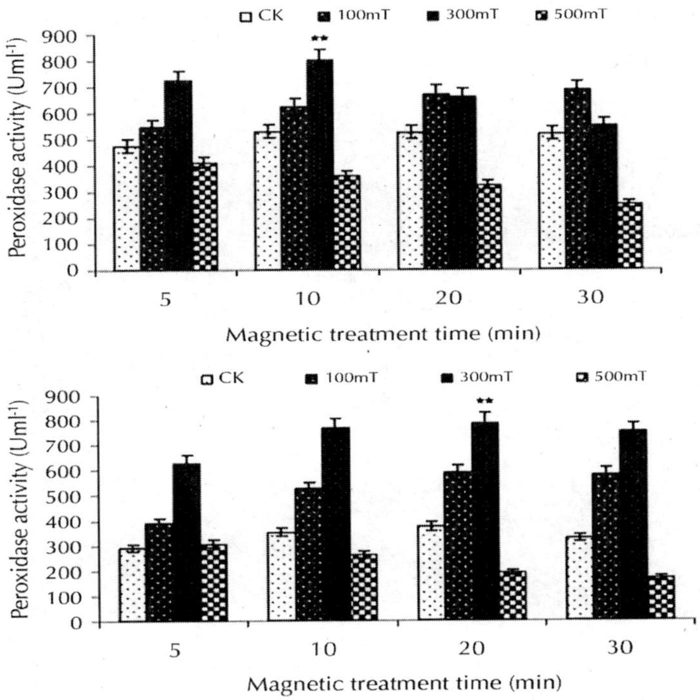

Figure 6: Effect of magnetic field on Peroxidase activity of*Paenibacillus*sp.I (upper) and*Paenibacillus*sp.II (lower). Error bars indicate SD among triplicates. Stars represent significance level at $P > 0.05$.

Figure 7 depicted the effect of different magnetic field on superoxide dismutase activity from*Paenibacillus* sp. Superoxide dismutase activity was found maximum at 20 min of magnetic treatment with intensity of 300 mT. This enzyme activity also showed the same trend as the previous ones (catalase and peroxidase activity) but here increased treatment time resulted increased superoxide dismutase activity. In all the three enzyme activities (catalase, peroxidase and superoxide dismutase) magnetic field intensity of 500 mT inhibited the enzyme activity because this magnetic intensity ruptured the cell shape which leads to decline in enzyme activities. The growth rate at 300 mT magnetic treatment was 340% and 153% for *Paenibacillus* sp. I and II respectively. So this study was in good agreement with Liu et al. (Liu et al. [1996]) who reported that appropriate magnetic field intensity enhanced the activities of hydrogen peroxidases, invertases, amylases and phosphatases in the three tested soils. Another study also revealed that magnetic field enhances the catalase and superoxide dismutase activity isolated from the roots of soybean (Celik et al. [2009]).

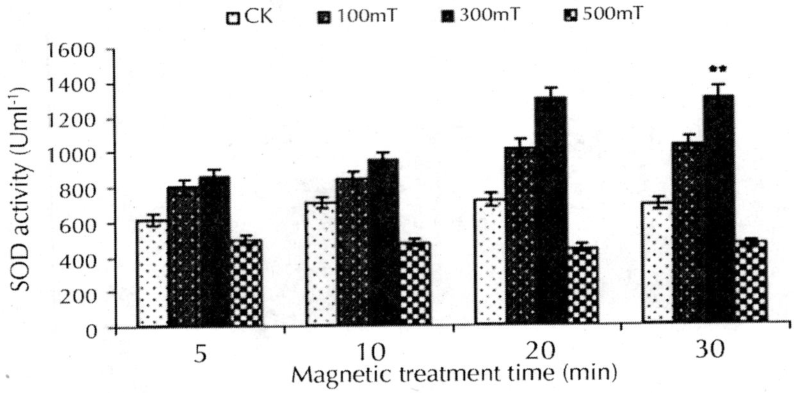

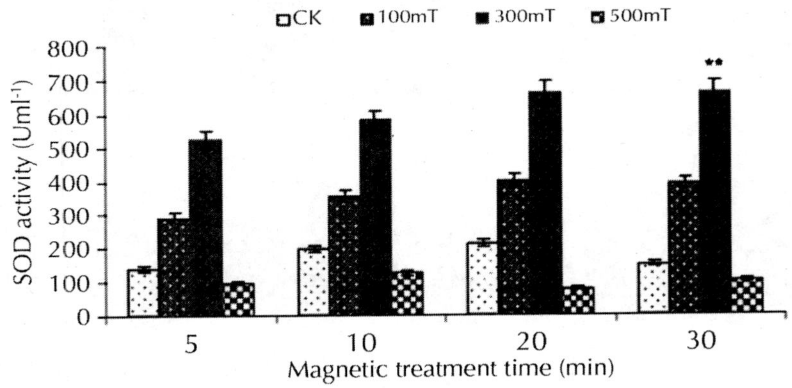

Figure 7: Effect of magnetic fields on Superoxide dismutase activity (SOD) of*Paenibacillus*sp.I (upper) and*Paenibacillus*sp.II (lower). Error bars indicate SD among triplicates. Stars represent significance level at $P > 0.05$.

CONCLUSIONS

In conclusion the magnetic treatment significantly enhances the bacterial population with shorter generation time. This increased population of *Paenibacillus* sp. would increase the nitrogen fixing efficiency thus leading to greater yield. The enzyme activities were also increased under the influence of magnetic treatment. Increased magnetic field intensity and longer magnetic processing time resulted ruptured bacterial cell which leads to cell death, thus reduction in nitrogen fixation efficiency. To achieve the better yield, appropriate magnetic field intensity and magnetic processing time is very important for this whole process.

AUTHORS' CONTRIBUTION

YY designed the study project, JL, XC performed experiments and collected data, DZ performed data analysis, and MI performed bioinformatics analysis and draft manuscript. All authors read and approved the final manuscript.

ACKNOWLEDGMENTS

This study was supported by the National Natural Science Foundation of China (Grant No. 40771111) and the Shenyang agricultural university youth fund (Grant No. 20070136).

REFERENCES

1. Agus JE, Steven DG, Brian ND (2000) Isolation and characterization of 2, 3-dichloro-Ipropanol-edgrading. Appl Environ Microbiol 66(7):2882-2887
2. Ash C, Priest FG, Collins MD (1993) Molecular identification of rRNA group 3 *Bacilli* (Ash, Farrow, Wallbanks and Collins) using a PCR probe test. Proposal for the creation of a new genus *Paenibacillus*. Antonie Van Leeuwenhoek 64:253-260
3. Ausubel M, Brent R, Kingston RE, Moore DD, Smit JA, Seidman JC, Struhl KS (1994) Current protocols in molecular biology, section 2.4. John Wiley and Sons, New York.
4. Baethgen WE, Alley MM (1989) A manual colorimetric procedure for measuring ammonium nitrogen in soil and plant. Commun Soil Sci Plant Anal 20(9&10):961-969
5. Barker AV, Bryson GM (2007) Nitrogen. In: Barker AV, Pilbeam DJ (eds) Handbook of Plant nutrition, CRC Press, Boca Raton. pp

21-50

6. Bergmeger H, Bergmeyer J, Grabl M (1983) Methods of enzymatic Analysis. Verlag Chemie Press, Weinheim.
7. Cai Y, Lai Z, Shen J (2006) Detection of superoxide dismutase activities in different organsim *Phyllanthus emblica* L. Chin J Trop Crops 27(4):29-33
8. Celik Ö, Büyükuslu N, Atak C, Rzakoulieva A (2009) Effects of magnetic field on activity of superoxide dismutase and catalase in *Glycine max* (L.) Merr. Roots. Pol J Environ Stud 18(2):175-182
9. Cheng X, Yi Y (2009) Effect of Magnetic Treatment on Amount and Generation Time of Slow-Growing Rhizobium (USDA110) and Fast-growing Rhizobium (USDA191). Southwest Chin J Agri Sci 22(5):1400-1403
10. Cheng X, Yi Y, and Du a (2007) Effects of Magnetic treatment on the Azotobacter contents in brown Earth. Chin J Soil Sci 38(5):1025-1027
11. Fadel MA, Wael SM, Mostafa RM (2003) Effect of 50 Hz, 0.2 mT magnetic fields on RBC properties and heart functions of albino rats. Bioelectromagnetics 24:535-545
12. Fang F, Li Y, Du GC (2004) Themo-alkali stable catalase *Themoascus aurantiacus* and its potential use in textile bleaching process. Chin J Biotechnol 20(3):423-428
13. Fisher RF, Long SR (1992) Rhizobium-plant signal exchange. Nature 357(6380):655-660
14. Gaafar EA, Hanafy MS, Tohamy EY, Ibrahim MH (2006) Stimulation and control of *E. coli* by using an extremely low frequency magnetic field. Rom J Biophys 16(4):283-296
15. Hagedorn F, Bucher JB, Schleppi P (2001) Contrasting dynamics of dissolved inorganic and organic nitrogen in soil and surface waters of forested catchments with Gleysols. Geoderma 100:173-192
16. He H, Zhu Y, Zhong K (1998) Effect of magnetic field on celhlase activity and conformation. J Jishou Univ 19(1):42-46
17. He H, Zhu Y, Fan Q (1999) the Effect of Magnetic Field on *Escherichia Coli* and Glutamic Acid Decarbolxylase. J Jishou

Univ 20(3):26-29

18. Hua H, Shen Y, Wu W (2008) Effects of magnetic field on seed quality, POD and SOD of Pinus massoniana. J Nanjing For Univ 32(3):39-42
19. Jia Y, Ma Y, Wang Z (2002) the enzymatic activity of tomato seeds with magnetic field treatment. Biotechnology 10(2):14-17
20. Jing Y, Zhang B, Wang Y, Lin X (1992) Effect of magnetic field on symbiotic nitrogen fixation of soybean nodules. Acta Botanica Sinica 34(5):364-368
21. Lal S, Tabacchioni S (2009) Ecology and biotechnological potential of *Paenibacillus*polymyxa: a minireview. Indian J Microbiol 49:2-10
22. Lamb C, Dixon RA (1997) the oxidative burst in plant disease resistance. Annu Rev Plant Physiol Plant Mol Biol 48:251-275
23. Lebuhn M, Heulin T, Hartmann A (1997) Production of auxin and other indolic and phenolic compounds by *Paenibacillus polymyxa* strains isolated from different proximity to plant roots. FEMS Micr Eco 22:325-334
24. Li J, Jiao Y, Yi Y (2007) Effects of magnetic field on catalase and peroxidase activities in brown earth. J Shenyang Agric Univ 38(1):70-74
25. Liu XY, Yi YL, Xia LH (1996) Effect of magnetic field on enzyme activities in main soils of Northeast China. Pedosphere 6(4):341-348
26. Liu YH, Xu LH, Tang X (2003) the Effect of Magnetic Field Treatment on Artificially Aging Pepper Seeds. J Shandong Inst Build Mat 17(3):286-288
27. Mao N, Huang Y, Zhang Z (2002) Study on Biological Effect of Magnetic Resonance and Magnetized Water on *Agricus bisporus* Strain 176. J Fujian Teachers Univ (Nat Sci) 18(3):61-65
28. Marschner H (1995) Mineral Nutrition of Higher Plants. Academic Press Limited, San Diego, London.
29. Mavingui P, Heulin T (1994) In vitro chitinase antifungal activity of a soil, rhizosphere and rhizoplane populations of *Bacillus polymyxa*. Soil Biol Biochem 26:801-803

30. McSpadden Gardener BB (2004) Ecology of *Bacillus* and *Paenibacillus* spp. in Agricultural Systems. Phytopathology 94:1252-1258

31. Mohamed AA, Ali FM, Gaafar EA, Magda HR (1997) Effects of magnetic field on the biophysical, biochemical properties and biological activity of Salmonella typhi., Master thesis submitted for Biophysics department, Faculty of science, Cairo University, Egypt.

32. Montes MJ, Mercade E, Bozal N, Guinea J (2004) *Paenibacillus antarcticus* sp. nov., a novel psychrotolerant organism from the Antarctic environment. Int J Syst Evol Microbiol 54:1521-1526

33. Ouyang J, Pei Z, Lutwick L, Dalal S, Yang L, Cassai N, Sandhu K, Hanna B, Wieczorek RL, Bluth M, Pincus MR (2008) Case report: *Paenibacillus thiaminolyticus*: a new cause of human infection, inducing bacteremia in a patient on hemodialysis. Ann Clin Lab Sci 38:393-400

34. Owen AG, Jones DL (2001) Competition for amino acids between wheat roots and rhizosphere microorganisms and the role of amino acids in plant N acquisition. Soil Biol Biochem 33:651-657

35. Pires MN, Seldin L (1997) Evaluation of Biology system for identification of strains of*Paenibacillus azotofixans*. Ant Van Leeu 71:195-200

36. Santos R, Herouart D, Sigaud S, Touati D, Puppo A (2001) Oxidative burst in alfalfa-Sinorhisobium meltloti symbiotic interaction. Mol Plant Micobe Interact 14(1):86-89

37. Sawada H, Kuykendall LD, Young JM (2003) Changing concepts in the systematics of bacterial nitrogen-fixing legume symbionts. J Gen Appl Microbiol 49(3):155-79|

38. Sierra S, Rodelas B, Martinez-Toledo MV, Pozo C, González-López J (1999) Production of B-group vitamins by two Rhizobium stains in chemically defined media. J Appl Microbiol 86:851-858

39. Vincent JM (1970) A Manual for the Practical Study of Root-Nodule Bacteria. Blackwell Scientific, Oxford.

40. Von der Weid I, Alviano DS, Santos ALS, Soares RMA, Alviano CS, Seldin L (2003) Antimicrobial activity of *Paenibacillus peoriae*

against a broad spectrum of phytopathogenic bacteria and fungi. J Appl Microbiol 95:1143-1151

41. Yan L, Zhu Y, He S, Cao Z (1997) Effect of static magnetic field on activity of immobilized a-amylase. Chinese Sci Bull 42(2):127-130
42. Zhu Y, Zhong K, He S (1996) Efect of Magnetic Field on Activity of Immobilized LDH. J Hunan Univ 23(5):57-61

Chapter

4

Does Overshoot in Leaf Development of Ponderosa Pine in Wet Years Leads to Bark Beetle Outbreaks on Fine-Textured Soils in Drier Years?

Wendy Peterman[1] and Richard H Waring[2]

[1]Conservation Biology Institute, 136 SW Washington Ave, Ste 202, Corvallis 97333, OR, USA

[2]Department of Forest Ecosystems and Society, Oregon State University, 321 Richardson Hall, Corvallis 97331, OR, USA

ABSTRACT

Background

Frequent outbreaks of insects and diseases have been recorded in the native forests of western North America during the last few decades, but the distribution of these outbreaks has been far from uniform. In some cases, recent climatic variations may explain some of this spatial variation along with the presence of expansive forests composed of dense, older trees. Forest managers and policy makers would benefit if areas especially prone to disturbance could be recognized so that mitigating actions could be taken.

Methods

We use two ponderosa pine-dominated sites in western Montana, U.S.A. to apply a modeling approach that couples information acquired via remote sensing, soil surveys, and local weather stations to assess where bark beetle outbreaks might first occur and why. Although there was a general downward trend in precipitation for both sites over the period between 1998 and 2010 (slope = −1.3, R^2 = 0.08), interannual variability was high. Some years showed large increases followed by sharp decreases. Both sites had similar topography and fire histories, but bark beetle activity occurred earlier (circa 2000 to 2001) and more severely on one site than on the other. The initial canopy density of the two sites was also similar, with leaf area indices ranging between 1.7-2.0 $m^2 \cdot m^{-2}$. We wondered if the difference in bark beetle activity was related to soils that were higher in clay content at site I than at site II. To assess this possibility, we applied a process-based stand growth model (3-PG) to analyze the data and evaluate the hypotheses.

Results

We found that when wet years were followed by drier years, the simulated annual wood production per unit of leaf area, a measure of tree vigor, dropped below a critical threshold on site I but not on site II.

Conclusion

We concluded that the difference in vulnerability of the two stands to beetle outbreaks can be explained largely by differences in gross photosynthesis attributed to the fact that an equivalent amount of stored water in the rooting zone (100 mm) is extracted less efficiently from fine-textured soils than from coarse-textured ones.

BACKGROUND

In the past few decades, world-wide forest decline has been reported in association with increases in temperature and decreases in precipitation (Allen et al. [2010]). Throughout forests in western North America the extent of disturbances caused by fire, insects and diseases is unprecedented (Raffa et al. [2008]; Bentz et al. [2009]). Drought reduces tree defenses, enabling insects to surpass a population threshold required for an outbreak (Berg et al. [2006]; Rouault et al. [2006]).

The impact of biotic agents is, however, expected to be spatially patchy (Dale et al. [2001]; Allen et al. [2010]). Density-dependent processes between the host organism and its attackers differ with local variations in soils and topography as well as with the physiological status of the host trees (Christiansen et al. [1987]; Allen et al. [2010]). Satellite-derived records of disturbance confirm that tree mortality is not uniformly distributed and that some of the variation can be attributed to differences in soil (Yoshiko and Mueller-Dombois [1995]; Brassier [1996]; Turner and Lambert [2005]; Bigler et al. [2006]; Hogg et al. [2008]; Harper et al. [2009]). In general, outbreaks are more prevalent on poorer quality sites and where there are large populations of overly-dense, older trees (Fettig et al. [2007]).

Furthermore, the mechanism by which beetles recognize, attack, and kill trees is well-described (Horntvedt et al. [1983]; Christiansen et al. [1987]; Franceschi et al. [2005]). Beetles bore into the phloem of the trees, creating tunnels in the secondary phloem and deposit their eggs, simultaneously allowing the entry of mutualistic fungi, which can disrupt the vascular system of the trees with their hyphae, leading to tree morality (Horntvedt et al. [1983]; Franceschi et al. [2005]; Raffa et al. [2008]). In healthy trees, a rapid flush of resin expels the initial beetle attack (Erbilgin et al. [2003]), and a barrier of dead or hardened tissue forms around the fungal infection to prevent it from spreading (Horntvedt et al. [1983]).

Because the beetles have co-evolved with their host tree species, they have developed mechanisms for overcoming tree defenses (Franceschi et al. [2005]). It is widely accepted that aggressive beetle species overwhelm the trees' physiological abilities to produce defensive tissues and chemicals by recruiting threshold numbers of individuals (Christiansen [1985]; Wallin and Raffa [2000], [2004]; Huber et al. [2004]; Franceschi et al. [2005]). Production of bark, resin ducts and toxins is very energetically costly and dependent on trees' ability to partition resources from local storage and photosynthesis to defense systems (Bryant and Julkunen-Tiitto [1995]; Franceschi et al. [2005]). The ability to create defensive mechanisms is compromised by water stress, temperature stress and pollution (Franceschi et al. [2005]; Raffa et al. [2008]).

Since the start of the 21st century, there are many places in the western United States where long-term patterns in precipitation have changed from past conditions (Allen et al. [2010]). The northern Rocky Mountain region is one of those places, as mapped by Waring et al. ([2011]). In western Montana, outbreaks of bark beetles on ponderosa pine (*Pinus ponderosa*) have been reported at lower elevations, but the pattern is not uniform (USFS FHTET [2010]). Such spatial variation in tree mortality raises a number of questions. Are trees more stressed on shallow soils than deeper soils in the region? Does soil texture play a role in drought stress for these trees? Is tree vigor somehow affected by interannual variations in precipitation? We hypothesize that 1) if tree photosynthesis is more constrained for trees growing on high clay soils than low clay soils under similar climate conditions, then beetle activity will be greater on clayey sites in dry years than on non-clayey sites, and 2) if there is an imbalance in interannual variations in leaf to stem growth ratios, then tree vigor will drop below threshold values for trees to become vulnerable to beetle attacks.

METHODS

Approach

We can discern climatic trends from published records and obtain maps of soil properties for areas in western Montana. It is difficult, however, to obtain ground measurements of plant water stress retrospectively or to assess subtle changes in canopy leaf area. Some researchers have correlated climatic variation from tree ring analyses (Garfinkel and Brubaker [1980]; Briffa et al. [1990]; Villalba [1990]) and from the analysis of variations in the stable isotopes of carbon (Panek and Waring [1997]). These types of analyses are spatially restricted and require a non-random approach to defining areas for sampling. In this paper, we propose an alternative approach by utilizing a process-based stand growth model driven by climatic variables that also incorporates important soil properties. Such models simulate the dynamics of trees as they shift their growth between leaves, stems and roots in response to seasonal and annual climatic and soil conditions.

In a stable environment, a forest canopy can be expected to reach a maximum leaf area index (LAI) that can be supported by available soil moisture and fertility (Waring [1983]). In western coniferous forests, LAI varies from < 1 to 12 $m^2 \cdot m^{-2}$ (Runyon et al. [1994]; Waring et al. [2014]). If growing conditions become less favorable, the LAI will decrease, often abruptly (Pook [1984]), which can result in increased tree mortality. Field experiments (Christiansen et al. [1987]) have shown that the ratio of annual wood production to LAI, termed "growth efficiency" (GE), has proven useful as a general index of tree vigor and specifically to identifying the vulnerability of pine stands to attacks from mountain pine beetles (*Dendroctonus ponderosae*) (Coops et al.[2009]).

We chose to use the 3-PG process-based growth model (Physiological Principles Predicting Growth) to simulate forest responses to differences in the climate and soil conditions that affect tree vigor and forest vulnerability to disturbances such as bark beetles. Process-based growth models use mathematical representations of biotic and abiotic processes to simulate components of both the water and carbon balances. 3-PG is widely used to represent the multiple interactions between climate, soils and forests with its main focus on predicting growth as measured by foresters (i.e. tree number, mean diameter, basal area, volume, with thinning and defoliation subroutines).

Study Areas

Using PRISM climate data in a geographic information system (GIS) environment, we selected two study sites in western Montana where climate conditions are consistent, but there have been uneven distributions of bark beetle activity. The area selected lies within the Lewis and Clark National Forest (Latitude: 47.1833° N, Longitude: 111.4500° W) and extends over an elevation range of 1400 m to 2900 m. The vegetation in the region is a mixture of evergreen conifers, subalpine fir (*Abies lasiocarpa*), Douglas-fir (*Pseudotsuga menziesii*), lodgepole pine (*Pinus contorta*), Englemann spruce (*Picea engelmannii*), and ponderosa pine (*Pinus ponderosa*). The climate is continental with minimum winter temperatures dropping to −12°C and summer maximums reaching 24°C. Precipitation averages between 90 and 100 $cm \cdot yr^{-1}$ with more in the spring than summer or winter. Both sites selected are classified as wilderness and are not used for timber production. Sixty-four percent of the ponderosa pine forest is considered "highly stocked" and susceptible to disease and insect attack (DeBlander [2002]). For our research, we selected a 24-year period between 1998 to 2010, when two major outbreaks of mountain pine beetle (*Dendroctonus ponderosae*) occurred, one circa 2001 to 2004, the other circa 2006 to 2010.

We chose two sites where beetle outbreaks were recorded and mapped by the USFS aerial surveys (Figure 1). The presence of ponderosa pines at these locations was verified using Forest Inventory and Analysis (FIA) plots and Little's tree range maps. A meteorological station was situated near site I at an elevation of 1844 m, where the mortality of ponderosa pine averaged 750 $trees \cdot ha^{-1}$ the first year of the beetle outbreak one circa 2001 and 950 $trees \cdot ha^{-1}$ at the beginning of the second outbreak circa 2007. Another meteorological station, located at an elevation of 2469 m, was near site II, where tree mortality was 400 $trees \cdot ha^{-1}$ circa 2001 and 1300 $trees \cdot ha^{-1}$ circa 2007. Forest canopies at both sites were fairly open with maximum leaf area indices ranging between 1.7 to 2.0 $m^2 \cdot m^{-2}$. The boundaries around site II were irregular in shape to include the full range of elevations at which ponderosa pine was recorded and where soil survey data were available. The dominant soils on site I are Mollisols, Entisols, Inceptisols and Vertisols developed on carbonate, alluvium and fine to medium-grained sedimentary and clastic parent materials. On site II, the dominant soils are Mollisols, Inceptisols, Alfisols and Entisols formed on carbonate, quartzite and coarse-grained volcanic parent materials.

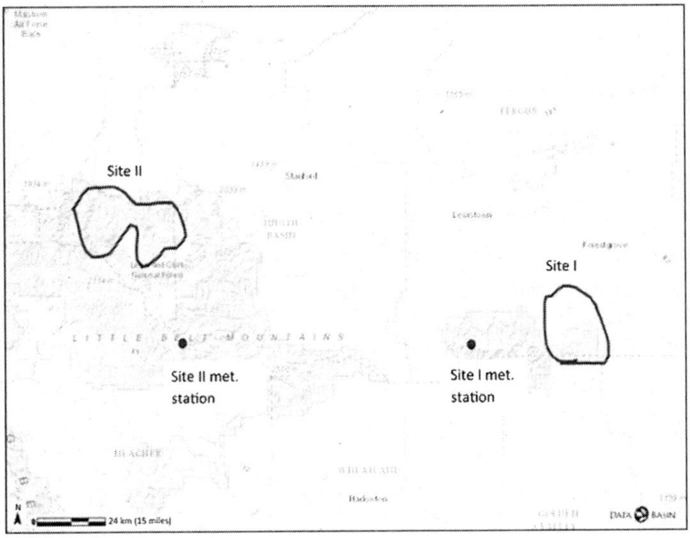

Figure 1: Study sites. Map showing two study sites in western Montana with black dots showing locations of meteorological stations used for climate data.

Climatic Conditions

Monthly meteorological data for local weather stations (Crystal Lake and Spur Park, MT) for the years 1985 to 2013 were downloaded in tabular format from the National Oceanic and Atmospheric Administration (NOAA) Climate Data Online website (NOAA and [2013]). Some missing values and errors in the temperature and precipitation tables were filled using correlations with the station at Lewistown, MT and the first-order automated surface observation station (ASOS) at the Helena, MT airport. Since solar radiation data were not available for these sites, we calculated monthly mean values, using differences between temperature extremes following procedures outlined by Coops et al. ([2000]). At each site, the correlation between calculated averaged solar radiation for each month of the year and that measured at Helena, MT was 0.99. Similarly, the evaporative demand, expressed as mean day-time vapor pressure deficit (VPD) each month and the number of frost days ($< -2°C$) were calculated from monthly mean temperature extremes using the 3-PG function tools (3PGpjs version 2.7, 3-PG version 1, September 2010).

GIS Analysis

In addition to climate conditions, soil characteristics such as water storage capacity, fertility and texture are well known to limit forest productivity (Viereck et al. [1983]; Powers et al. [2005]; Xu et al. [2008]; Puhlick et al. [2012]). Initial comparisons of forest mortality for all tree species in western Montana and soil characteristics were made by intersecting polygons in the Soil Survey Geographic database (SSURGO) soil data (NRCS [2014]; min. 1:24000) with polygons in the annual forest die-off data (USFS FHTET, min. 1:100000) from 1998 to 2010 to join the attributes from the soil map unit polygons to the tree mortality polygons. The US Forest Service (USFS) aerial surveys combined low-level flights (315 to 630 m above ground level) and US Geological Survey paper maps (1:100000 scale), as well as a digital sketchmap system (GPS and GIS database) to record the area of disturbance on an annual basis. Due to the size of the polygons and the patchy nature of forest insect activity, some polygons may have contained unaffected areas. For the sake of recording new damage, only trees with yellow, brown or red foliage or some defoliation were mapped as part of the aerial survey. Dead trees with no foliage were assumed to be recorded in a previous year's study. Soil mapping units were aggregated to generalize common properties and to minimize differences associated with variable spatial scales. Tree mortality data was summarized by percent of the area of dead trees and compared to the percent of the area where tree cover was greater than 50% (Fry et al. [2011]) and no mortality was recorded during the time period. To refine this study to specific sites and a specific tree species, soil characteristics between two areas in Montana were then compared to seek possible differences among ponderosa pine forests where tree mortality was recorded between 1998 and 2010. GIS analysis was performed by overlaying datasets representing ponderosa pines killed by mountain pine beetles at the two sites between 1998 and 2010 (Meddens et al. [2012]) along with SSURGO soils data.

Remote Sensing

Landsat graphs of monthly greenness (NDVI) values for the years 1998 to 2010 were downloaded from the Glovis Data Viewer (Tile IDs: LT50380271998231PAC00, path 38, row 27, product: ETM + L1T and LT50390271999193XXX02 path 39, row 27, product: ETM + L1T) (USGS [2014]). These were used to compare trends in the normalized difference vegetation indices (NDVI) (Eq. 1) for both sites and to calculate estimates for maximum annual leaf area index (LAI_{max}). Estimates were made from a correlation established between NDVI and LAI reported by Turner et al. ([1999]):

$$NDVI = 0.5724 + 0.0989LAI - 0.0114LAI^2 + 0.0004LAI^3, R^2 = 0.74, SEE = 0.14 \quad (1)$$

This method was chosen due to the similarity in the range of NDVI and LAI values between their sites and ours. Additionally, the testing of their correlations over a broad range of temperate sites using Landsat 5 surface reflectance data with atmospheric correction gave us confidence that it would be applicable in our study area.

3-PG Process Model

The 3-PG model uses monthly averages of climatic variables to calculate a water balance, carbon budget and energy balance that is summed for the year. At a minimum, precipitation and maximum and minimum air temperature must be provided from meteorological stations. Other climate variables can be calculated using the 3-PG function tools and verified using correlations with nearby weather station records. Four soil variables are also included in the model: minimum and maximum available soil water storage capacity (ASW) in the top 1500 mm of soil, soil fertility (FR) and soil texture. The soil receives a fraction of the precipitation for each month, and the soil water that is not released through evapotranspiration is added to the next month's water balance or allowed to run off (Landsberg and Waring [1997]; Waring et al.

[2014]). ASW is defined as the difference between field capacity (soil moisture held at 1/3 bar tension) and wilting point (soil moisture held at 15 bar tension). FR in the model is a rating on a scale of 0 to 1, with 0 being the lowest and 1 being the highest. For this study, we fixed the photosynthetic capacity at 0.05 mol C per mol photon (2.75 $gC \cdot MJ^{-1}$ of absorbed photosynthetically-active radiation). This value has been measured and used in other ponderosa pine growth modeling exercises (Waring et al. [2014]), and sensitivity analysis of the model showed that lower photosynthetic capacity settings required unrealistically high FR rankings. In the model, photosynthesis is restricted by frost, high vapor pressure deficit, suboptimal temperatures and soil water deficits. Each month, the upper limits on gross photosynthesis are set by the amount of light (PAR) absorbed by the LAI. Approximately half the total gross photosynthesis is assumed to be lost through plant respiration, and the rest is available for growth. Growth is partitioned above- and below-ground as a function of soil fertility and to a lesser extent, soil water deficit (Landsberg and Waring [1997]).

Model Parameterization and Assumptions

For this modeling exercise, we used the Excel version of 3-PG with growth parameters for ponderosa pine based on available literature and Landsat-derived LAI (Table 1). The model parameters can be set to accommodate selected silvicultural treatments such as thinning and fertilization, but these options were not used here. Sensitivity analyses for three soil variables, maximum ASW, FR and soil texture were performed by running the model with two soil variables held constant while the third was changed incrementally. Annual outputs for gross photosynthesis or gross primary productivity (GPP), LAI and Penman-Monteith transpiration were compared to evaluate the effects of each soil variable on key physiological responses. Once realistic values for soil parameters and LAI were determined through GIS, remote sensing and sensitivity analyses, soil texture, FR and ASW were set at values that constrained simulated LAI to the estimated maximum value (1.7 to 2.0 $m^2 \cdot m^{-2}$). Model output was compared at annual intervals to interpret the simulated effects of climatic variation on stand growth and water balances for each site.

Table 1: Summary of parameter settings used in the 3-PG simulations for ponderosa pine

3-PG Parameter	Units	Ponderosa pine
Biomass partitioning and turnover		
Allometric relationships & partitioning		
Foliage: stem partitioning ratio @ D = 2 cm	–	1.328
Foliage: stem partitioning ratio @ D = 20 cm	–	0.75
Constant in the stem mass v. diam. relationship	–	0.0046
Power in the stem mass v. diam. relationship	–	2.97
Maximum fraction of NPP to roots	–	0.8
Minimum fraction of NPP to roots	–	0.25
Litterfall & root turnover		
Maximum litterfall rate	$1 \cdot month^{-1}$	0.021
Litterfall rate at t = 0	$1 \cdot month^{-1}$	0.001
Age at which litterfall rate has median value	months	36
Average monthly root turnover rate	$1 \cdot month^{-1}$	0.015
NPP & conductance modifiers		
Temperature modifier (fT)		
Minimum temperature for growth	°C	−7
Optimum temperature for growth	°C	18
Maximum temperature for growth	°C	40
Frost modifier (fFRost)		
Days production lost per frost day	Days	1
Soil water modifier (fSW)		
Moisture ratio deficit for f_q = 0.5	–	0.7
Power of moisture ratio deficit	–	9
Canopy structure and processes		
Specific leaf area		
Specific leaf area at age 0	$m^2 \cdot kg^{-1}$	3.1
Specific leaf area for mature leaves	$m^2 \cdot kg^{-1}$	3.1
Age at which specific leaf area = $(SLA_0 + SLA_1)/2$	Years	2.5
Light interception		
Extinction coefficient for absorption of PAR by canopy	–	0.5
Age at canopy cover	Years	15

Maximum proportion of rainfall evaporated from canopy	–	0.1
LAI for maximum rainfall interception	–	5
Production and respiration		
Canopy photosynthetic capacity	mol C·mol PAR^{-1}	0.05
Ratio NPP/GPP	–	0.47
Conductance		
Minimum canopy conductance	m·s^{-1}	0
Maximum canopy conductance	m·s^{-1}	0.016
LAI for maximum canopy conductance	–	5
Defines stomatal response to VPD	1·mBar^{-1}	0.05
Canopy boundary layer conductance	m·s^{-1}	0.14

Peterman and Waring

Peterman and Waring Forest Ecosystems 2014 1:24, doi: 10.1186/s40663-014-0024-1

It was assumed for this exercise that ponderosa pines attacked by bark beetles were more than 60 years old (Coops et al. [2009]) and that stand densities exceeded 18 m^2·ha^{-1} (Schmid and Mata [1992], [2005]; Negrón and Popp [2004]), a threshold above which ponderosa pine stands in the Rocky Mountains have been shown to be highly susceptible to beetle attacks (Negrón and Popp [2004]; Zausen et al. [2005]). These values were collected from the literature, because local resources only stated that ponderosa pines on the study sites were of ages and densities that made them susceptible to bark beetle attacks (DeBlander [2002]).

RESULTS

Climate Data Trends

Warmer, drier trends in climate over the twenty-four year period of this study signaled that atmospheric conditions were becoming more conducive to forest vulnerability to water stress and disturbances (Figure 2). Temperature trends were upward on both sites, with warmer conditions recorded near site I than site II (Figure 2A). Monthly precipitation, when averaged for each year, showed a slight decrease over the twenty-four year period from 1989 to 2013 (Figure 2B).

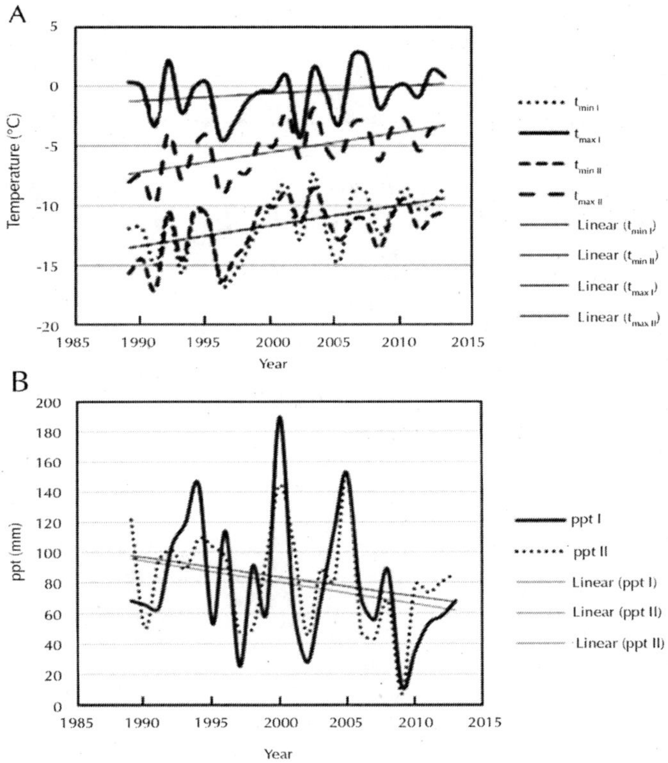

Figure 2: Climate trends. Average annual minimum and maximum (A) temperatures and (B) precipitation for site I and site II between the years 1985 and 2013.

Soil Sensitivity Analysis for Evaluating Soil Parameters to Constrain LAI

The relative influence of soil water storage, soil fertility, and soil texture varied in their effects on GPP at the two sites. On both sites, values of ASW ≤ 100 mm caused a reduction in simulated productivity in years with below average precipitation. Likewise on both sites, modeled increases in soil fertility (FR) caused an increase in GPP, resulting in the production of more leaf area, but as FR approached 0.5, the relative effects of increasing FR decreased. During periods of lower precipitation, the modeled difference in FR effect was less. The influence of soil texture on modeled tree growth was most evident on site I, particularly between the years 2001–2005 and 2006–2009 (Figure 3).

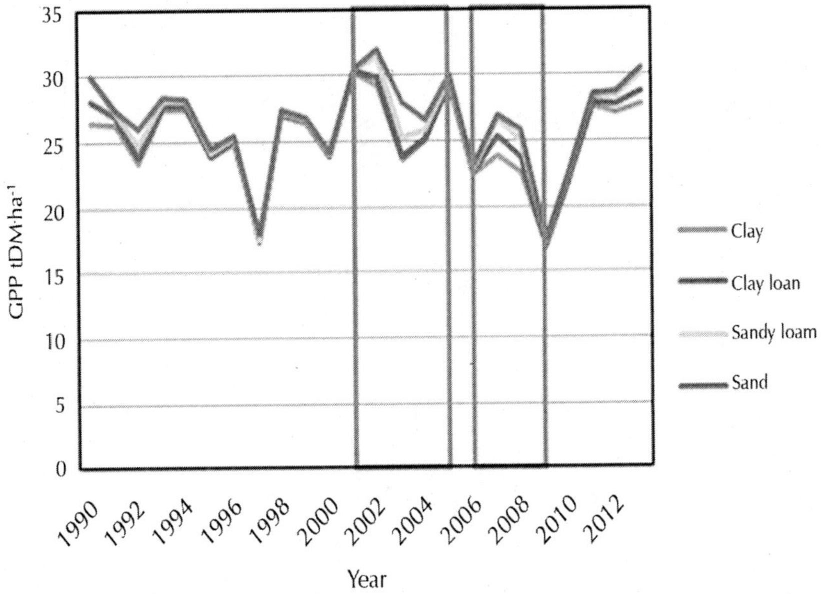

Figure 3: Modeled gross photosynthesis with four different soil textures. Changes greater than 10% are considered significant. In this graph, the difference between sand and clay is 14.3% between 2001 and 2005 and 11.5% between 2006 and 2009. Gray boxes highlight the periods of greater influence of soil texture.

GIS Analysis of Soil Characteristics and Forest Mortality Areas

Initial GIS analysis of all tree species in western Montana compared to soil texture class showed that 99% of the area with mortality between 1998 and 2010 had clayey soils, while 94% of the area with no mortality had loam to sandy loam soils (Figure 4), and 5% of the area of tree persistence was on clayey soils. Furthermore, GIS analysis of soil characteristics on the two sites in Montana showed that soil water storage capacities varied considerably across the landscape. Most soils in the region could store between 50 and 150 mm of water. Sand content was generally higher on site II, but the main difference between the soils was the clay content. Clay content on site I ranged between 22% and 60%, with 80% of the total area containing soils with between 27% to 37% clay content. The clay content of soils at site II was between 0 and 28%, and for 90% of the area, it was 18% to 25%. A graph of clay content versus the percent of the area of trees killed by mountain pine beetles showed a close relationship between the area affected and the clay content of the soil (Figure 5). The percent of the total area of beetle-killed trees on soils with > 20% clay content was approximately 94% on site I and 65% on site II.

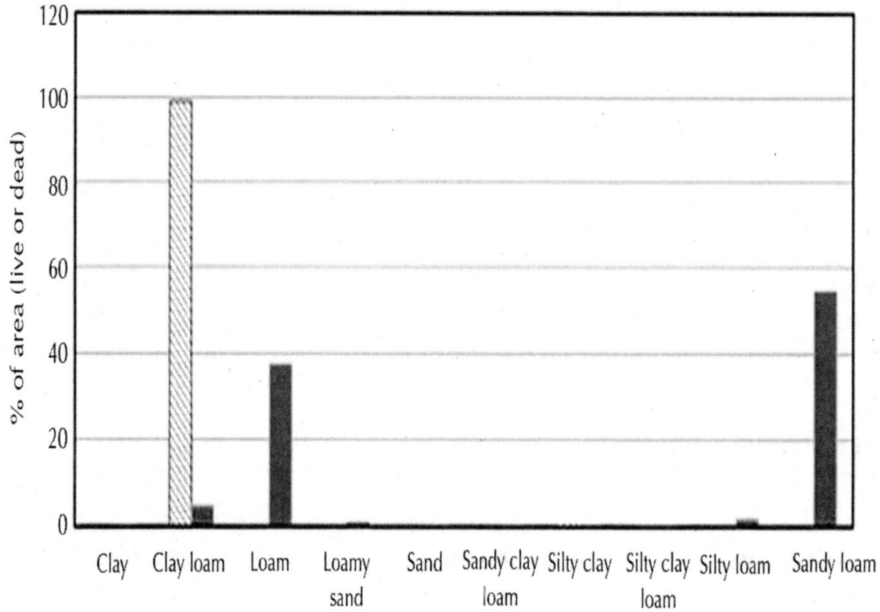

Figure 4: Soil texture and tree mortality in western Montana. This graph shows the percent of the area of tree mortality or tree persistence on each soil texture type for all tree species in all of western Montana, USA between 1998 and 2010. Legend: Striped fill = killed trees, solid fill = live trees.

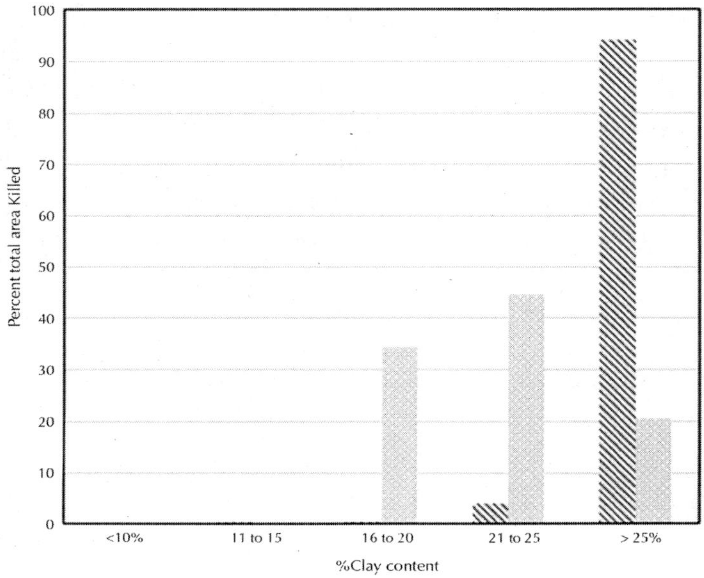

Figure 5: Site specific clay content and tree mortality. The graph shows percent clay content of the soil versus the percent of the total area of pines killed on each site. Legend: striped fill = site I mortality, dotted fill = site II mortality.

Modeled Results

Inter-annual variation in simulated LAI showed large increases corresponding with wetter periods and decreases during drier periods (Figure 6). On site I, there was a higher LAI between 2003 and 2006, followed by a reduction in LAI circa 2007 to 2010 (not including the thinning attributed to beetle-caused mortality). On site II, simulated LAI steadily increased between 2000 and 2005 to a peak of 1.8 $m^2 \cdot m^{-2}$ and then decreased between 2005 and 2010.

Figure 6: Simulated leaf area index. The graphs show simulated LAI between 1998 and 2010 for site I (A) and site II (B). Gray boxes highlight the periods of high beetle activity on each site.

Figure 7 shows growth efficiency (GE = stem growth/LAI × 100), normalized over the entire modeled time period. On site I, modeled GE was 10% below average during outbreak one, and it was approximately 25% below average during outbreak two. Between outbreaks, however,

GE was approximately 20% above average. On site II, modeled GE was above average preceding outbreak one, at which point it dropped to the average, and then went above average until outbreak two, when it dropped to approximately 15% below the average. Another difference between the two graphs was that the variance of GE on site I was 0.014, and on site II, it was 0.008, reflecting larger departures from the mean on site I.

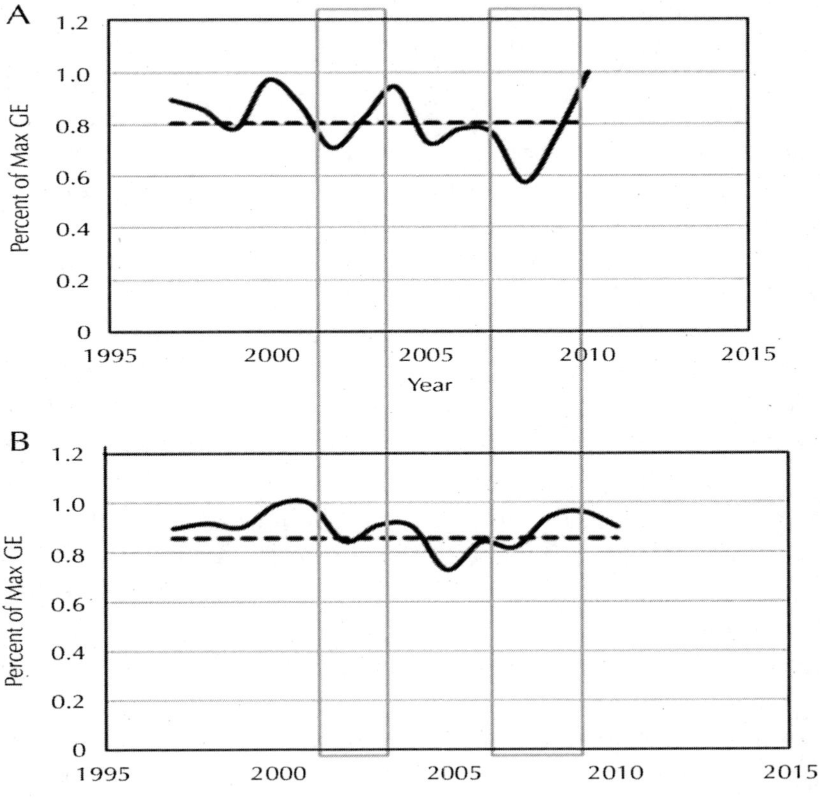

Figure 7: Growth efficiency comparison. The graphs show growth efficiency (GE) normalized over the modeled average for the time period for (A) site I and (B) site II. Gray boxes highlight the time periods of high beetle activity on each site. Legend: solid black lines = annual growth efficiency, dashed black lines = average growth efficiency for the entire time period.

DISCUSSION

Simulated Sensitivity of Ponderosa Pine Growth to Soil and Climate

In the sensitivity analysis, simulated effects of soil texture on GPP were evident on both sites at the end of the study period, but the importance of soil texture was particularly pronounced on site I between 2001 to 2005 and 2006 to 2009. Therefore, it may be that the amount of rain or snowmelt that actually reached the available root-zone water supply during drier periods was lower at site I than site II, because of differences in percolation rates between coarse and fine-textured soils. Water in coarse soils is acted upon only by gravitational forces, and therefore it travels easily through the soil profile and is readily extracted by tree roots (Hacke et al. [2000]). Conversely, water in clayey soils is held by capillary forces of cohesion and adhesion, which make drainage to deeper soil layers slower and uptake by tree roots more difficult. We propose that when there is adequate precipitation to allow deep drainage, trees can survive on clayey soils, however, in periods of lower precipitation, there may not be enough water reaching the rooting zone to meet the evaporative demand. Wu and Chen ([2013]) showed that inter-annual decreases in soil moisture had more influence on forest productivity than did increased summer temperatures, and Nambiar and Sands ([1992]) showed that the high bulk density and the increased tension per unit of water extracted from clayey soils limited transpiration. Our modeling exercise suggests that even with the same available soil water storage capacity, coarse-textured soils provide less constraint on simulated GPP and tree growth than fine-textured soils.

In ponderosa pine ecosystems, soil texture is an important factor in forest water stress during lower precipitation years and post-disturbance regeneration (Scianna [2011]). Other ecologists have observed greater pine productivity on coarse-textured soils than fine-textured soils (Franklin and Dyrness [1969]; Hoffman and Alexander [1976]), and presettlement ponderosa pine growth was denser on coarse-textured soils (Covington and Moore, [1994]). Puhlick et al. ([2012]) showed that the survival of ponderosa pine seedlings in the southwestern USA was highly dependent on soil texture and parent material. In basalt-derived soils, the water available for ponderosa pine use becomes limited at a moisture content of 10% (Heidmann and King [1992]), whereas in coarse-textured soils, developed from sedimentary rocks, water is not limiting to ponderosa pines until it reaches 1.5% moisture content.

Simulated Effects of Interannual Variations in Precipitation

An interesting result of this study is the simulated response of trees to the inter-annual variations in precipitation. In the simulations, the trees added foliage and increased LAI during the years with above-average precipitation. Could this response potentially have adverse consequences if the following year or multiple years have below-average precipitation, and stem growth is unable to meet the evaporative demand of the foliage? Samuelson et al. ([2004]) showed a lag effect in the response of foliage to large changes in precipitation by correlating current growth of loblolly pine (*Pinus taeda*) with the previous year's LAI, and Sperry et al. ([2002]) explained how combinations of soil texture and xylem can impose "hydraulic limits" on the ability of the foliage to maintain adequate pressure potentials to support a continuous water column. In our simulations, when precipitation fell below average, the trees quickly shed foliage to adapt to the drier conditions, and tree vigor was negatively affected by the sudden drop in leaf area as indicated by the decrease in the following year's GE (ratio of change in the change in LAI to stem diameter). Wright et al. ([1979], [1984]) reported that severe decreases in foliage can make forests more vulnerable to disease. Furthermore, previous studies have shown when GE drops below a threshold level, tree resistance to disease is greatly compromised (Larsson et al.[1983]; Waring and Pitman [1983]). In this study, simulated GE did not drop below average during outbreak 1 on site II, but it did drop below average prior to outbreak 2. On site I, however, simulated GE dropped below average prior to both outbreaks. The drop in GE below average signals a decrease in tree vigor and a possible imbalance between stem growth and leaf area. This is not conclusive evidence of an overshoot in leaf development that stressed the trees in drier years but it does give some support to that hypothesis that an imbalance in leaf to stem growth ratios reduces tree vigor, making trees more vulnerable to disturbance. Tree ring and carbon isotope analysis of a sample of trees on the study sites would give measurable data for evaluating this question, which would be a good field study to follow this modeling exercise.

Data and Model Limitations

All data are subject to human error and limitations on their ability to accurately and realistically represent nature, based on the spatial scale and precision of measurements taken. According to the USDA NRCS, Soil Survey Division Staff ([1995]), the US General soil map (STATSGO) is made by generalizing more detailed maps and is not suitable for local planning or productivity analyses. Semi-detailed maps like SSURGO, are based on field measurements, aerial photography and visual inspection of landscape features. They are considered suitable for determining differences and similarities in the soil characteristics of land areas greater than approximately 1 ha and can be used for planning on the level of farms and forests. They are also recommended for use in modeling with other environmental datasets representing geology, vegetation, elevation and climate. SSURGO soil maps are not meant to be used in place of onsite sampling for specific locations and uses. Soils grouped into one map unit will have a great deal of variation in reality, and land use practices will alter soil characteristics from their originally mapped condition (USDA NRCS, Soil Survey Division Staff [1993]). Therefore, it is imperative to make detailed measurements on specific plots for field research studies and monitoring programs.

There are also considerations when using climate data. As with soil data, the area represented by a given weather station is also subject to spatial variation due to changes in local topography. Equipment from meteorological stations is subject to malfunction and operational differences that bias measurements (NCAR [2014]). For example, intense rainfall and strong winds can affect the amount of precipitation measured in collectors. Comparing monthly values with nearby stations to check for correlations can help to adjust for errors and missing values.

Remote sensing satellites are limited by spatial, temporal and spectral resolutions, dynamic range and interference by clouds, snow or aerosols. Remote sensing data may not be useful in estimating

LAImax for densely-forested regions, because satellite sensors saturate from the high reflectance (Zhao et al. [2010]; Waring et al. [2014]). Comparing greenness values for several years can help analysts estimate reasonable values for LAImax for model calibration, but disturbances may be difficult to ascertain due to coarse spatial or spectral resolution or cloud interference (Hilker et al. [2009]). Atmospheric interference such as cloud cover causes more visible light to be reflected back to the sensor than would be the case if the light reached the surface and was absorbed by vegetation. When this occurs, the difference between the infrared reflectance and the visible red reflectance, used in calculating NDVI, is smaller than it would be if the emitted radiation were intercepting vegetation. This can result in unrealistic values of NDVI and thus calculated LAI. In the event of an equipment malfunction, data may be missing for multiple months (Landsat [2014]).

In addition to awareness of the limitations of data used in modeling, acknowledgment of the limitations of the models themselves is also important. In this exercise, we used the Excel version of 3-PG with climate data from local weather stations, and thus the simulation results only apply locally. New simulations would need to be run to apply this approach to other sites with appropriate soil and climate values. In addition, 3-PG is run on a monthly time-step, so it estimates general trends in annual carbon, energy and water balances, and it does not have specific algorithms to incorporate the effects of snowmelt dynamics on tree physiology. There is a general assumption that precipitation falling on days with temperatures below −2°C is received as snow (Waring et al.[2014]). Although 3-PG has some subroutines to simulate the effects of disease on forest productivity, it does not simulate the population dynamics of beetles or their responses to climate change.

CONCLUSIONS

GIS analysis of where tree mortality and bark beetle infestations occurred in western Montana showed that approximately 94% of the mortality area on site I and 65% of the mortality area on site II co-occurred with high clay soils (>20% clay content). Sensitivity analysis in the 3-PG model showed that the effect of soil clay content on photosynthesis was more pronounced during drier years. Simulations also showed there may have been an effect of large interannual variations in precipitation, causing an imbalance between leaf area and stem growth, with negative impacts on tree vigor. The combination of water stress and reduced tree vigor could have made the trees more vulnerable to beetles and their associated fungi.

Outbreaks of native insects and diseases are expected to occur in response to future warming temperatures and a drier environment, but tree mortality is not likely to be uniformly distributed across the landscape. Forest managers and policy makers can benefit from methods to help identify where forests are most susceptible to disturbance, so they might ameliorate conditions where and when it will do the most good. Retrospective analyses as performed in this modeling exercise suggest that two specific site variables would be most useful to map in predicting ponderosa pine disturbances: inter-annual variations in canopy leaf area index and soil texture.

AUTHORS' CONTRIBUTIONS

WP performed 3-PG modeling and sensitivity analyses, all GIS analyses, climate correlations, and remote sensing data analyses and data preparation. RW provided text, interpretation and expert supervision of all analyses. Both authors read and approved the final manuscript.

ACKNOWLEDGMENTS

We would like to thank Thomas Hilker (Oregon State University), Paul Adams (Oregon State University), Ron Reuter (Oregon State University), and Dominique Bachelet (Conservation Biology Institute) for their careful and insightful editing and feedback in writing this article. We thank Brendan Ward (Conservation Biology Insitute) for the use of his Python script to help extract and aggregate soil map units and their attributes from the NRCS soil survey maps.

REFERENCES

1. Allen CD, Macalady AK, Chenchouni H, Bachelet D, McDowell N, Vennetier M, Kitzberger T, Rigling A, Breshears DD, Hogg EH,

Gonzalez P, Fensham R, Zhang Z, Castro J, Demidova N, Lim J-H, Allard G, Running SW, Semerci A, Cobb N: A global overview of drought and heat-induced tree mortality reveals emerging climate change risks for forests. *Forest Ecol Manag* 2010, 259:660-684.

2. Bentz BJ, Regniere J, Fettig CJ, Hansen EM, Hayes JL, Hicke JA, Kelsey RG, Negron JF, Seybold SJ: Bark beetle outbreaks in Western North America: causes and consequences. *Bioscience* 2009, 60:602-613.
3. Berg EE, Henry JD, Fastie CL, DeVolder AD, Matsuoka SM: Spruce beetle outbreaks on the Kenai Peninsula, Alaska, and Kluane National Park and Reserve, Yukon Territory: Relationship to summer temperatures and regional differences in disturbance regimes. *Forest Ecol Manag* 2006, 227:219-232.
4. Bigler C, Bräker OU, Bugmann H, Dobbertin M, Rigling A: Drought as an inciting mortality factor in Scots pine stands of the Valais, Switzerland. *Ecosystems* 2006, 9:330-343.
5. Brasier CM: *Phytophthera cinnamomi* and oak decline in southern Europe. Environmental constraints including climate change. *Ann for Sci* 1996, 53:347-358.
6. Briffa KR, Bartholin JS, Eckstein D, Jones PD, Karlén W, Schweingruber FH, Zetterberg P: A 1,400-year tree-ring record of summer temperatures in Fennoscandia. *Nature* 1990, 346:434-439.
7. Bryant JP, Julkunen-Tiitto R: Ontogenic development of chemical defense by seedling resin birch: Energy cost of defense production. *J Chem Ecol* 1995, 21:883-896.
8. Christiansen E: Ips/Ceratocystis – infection of Norway spruce: what is a lethal dosage? *Z Angew Entomol* 1985, 99:6-11.
9. Christiansen E, Waring RH, Berryman AA: Resistance of conifers to bark beetle attack: Search for general relationships. *For Ecol Manag* 1987, 22:89-106.
10. Coops NC, Waring RH, Moncrieff J: Estimating mean monthly incident solar radiation on horizontal and inclined slopes from mean monthly temperature extremes. *J Biometeorol* 2000, 44:204-211.
11. Coops NC, Waring RH, Wulder MA, White JC: Prediction and assessment of bark beetle-induced mortality of lodgepole pine

using estimates of stand vigor derived from remotely sensed data. *Remote Sens Environ* 2009, 12:1058-1066.

12. Covington WW, Moore MM: Postsettlement changes in natural fire regime and forest structure: ecololigical restoration of old-growth ponderosa pine forests. *J Sustain Forestry* 1994, 2:153-181.
13. Dale VH, Joyce JA, McNulty S, Neilson RP, Ayres MP, Flannagan MD, Hanson PJ, Irland LC, Lugo AE, Peterson CJ, Simberloff D, Swanson FJ, Stocks BJ, Wotton BM: Climate change and forest disturbances. *Bioscience* 2001, 51:723-734.
14. DeBlander LT: *Forest resources of the Lewis and Clark National Forest. USDA Forest Service Rocky Mountain Research Station*. Ogden, Utah; 2002.
15. Erbilgin N, Powell J, Raffa K: Effect of varying monoterpene concentrations on the response of*Ips pini* (Coleoptera: Scolytidae) to its aggregation pheromone: implications for pest management and ecology of bark beetles. *Agr For Entomol* 2003, 5:269-274.
16. Fettig CJ, Klepsig KD, Billings RF, Munson AS, Nebeker TE, Negron JF, Nowak JT: The effectiveness of vegetation management practices for prevention and control of bark beetle infestations in coniferous forests of the western and southern United States. *For Ecol Manag* 2007, 238:24-53.
17. Franceschi VR, Krokene P, Christiansen E, Krekling T: Anatomical and chemical defenses of conifer bark against bark beetles and other pests. *New Phytol* 2005, 167:353-375.
18. Franklin JF, Dyrness CT: *Vegetation of Oregon and Washington. USDA Forest Service, Research Paper PNW-80*. Pacific Northwest Forest and Range Experiment Station, Portland, OR; 1969.
19. Fry J, Xian G, Jin S, Dewitz J, Homer C, Yang L, Barnes C, Herold N, Wickham J: Completion of the 2006 National Land Cover Database for the Conterminous United States. *PE&RS* 2011, 77:858-864.
20. Garfinkel HL, Brubaker LB: Modern climate-tree growth relationships and climatic reconstruction in subarctic Alaska. *Nature* 1980, 286:872-874.
21. Hacke UG, Sperry JS, Ewers BE, Ellsworth DS, and Schaefer KVR, Oren R: Influence of soil porosity on water use in *Pinus taeda*. *Oecologia* 2000, 124:495-505.

22. Harper RJ, Smettem KRJ, Carter JO, McGrath JF: Drought deaths in *Eucalyptus globulus* (Labill.) plantations in relation to soils, geomorphology and climate. *Plant Soil* 2009, 324:199-207.

23. Heidmann LJ, King RM: *Effect of prolonged drought on water relations of ponderosa pine seedlings grown in basalt and sedimentary soils. USDA For Serv Res Paper RM-301*. Rocky Mountain Forest and Range Experiment Station, Fort Collins, CO; 1992.

24. Hilker T, Coops NC, Coggins SB, Wulder MA, Brown M, Black TA, Nesic Z, Lessarde D: Detection of foliage conditions and disturbance from multi-angular high spectral resolution remote sensing. *Remote Sens Environ* 2009, 113:421-434.

25. Hoffman GR, Alexander RR: *Forest vegetation of the Big Horn Mountains, Wyoming, a habitat type classification. USDA Forest Service, Research Paper RM-170*. Rocky Mountain Forest and Range Experiment Station, Fort Collins, CO; 1976.

26. Hogg EH, Brandt JP, Michaelian M: Impacts of a regional drought on the productivity, dieback and biomass of western Canadian aspen forests. *Can J Forest Res* 2008, 38(6):1373-1384.

27. Horntvedt R, Christiansen E, Solheim H, Wang S: Artificial inoculation with *Ips typographus*-associated blue-stain fungi can kill healthy Norway spruce trees. *Medd Nor Inst Skogforsk* 1983, 38:1-20.

28. Huber DPW, Ralph S, Bohlmann J: Genomic hardwiring and phenotypic plasticity of terpenoid-based defenses in conifers. *J Chem Ecol* 2004, 30:2399-2418. |

29. Landsat: *Landsat Missions*. 2014.

30. Landsberg JJ, Waring RH: A generalized model of forest productivity using simplified concepts of radiation-use efficiency, carbon balance and partitioning. *For Ecol Manag* 1997, 95:209-228.

31. Larsson S, Oren R, Waring RH, Barrett JW: Attacks of mountain pine beetle as related to tree vigor of ponderosa pine. *Forest Sci* 1983, 29:395-402.

32. Meddens AJH, Hicke JA, Ferguson CA: Spatial and temporal patterns of observed bark beetle-caused tree mortality in British Columbia and the western. *US Ecol Appl* 2012, 22:1876-1891.

33. Nambiar EKS, Sands R: Effects of compaction and simulated root channels in the subsoil on root development, water uptake and growth of radiate pine. *Tree Physiol* 1992, 10:297-306.
34. NCAR: *Climate Data Guide*. 2014.
35. Negrón JF, Popp JB: Probability of ponderosa pine infestation by mountain pine beetle in the Colorado Front Range. *For Ecol Manag* 2004, 191:17-27.
36. NOAA: *National Climate Data Center: Climate Data Online*. 2013.
37. NRCS: *Web Soil Survey*. 2014.
38. Panek JA, Waring RH: Stable carbon isotopes as indicators of limitation to forest growth imposed by climate stress. *Ecol Appl* 1997, 7:854-863.
39. Pook EW: Canopy dynamics of *Eucalyptus maculata* Hook. II. Canopy leaf area balance. *Aust J Bot* 1984, 32:405-413.
40. Powers RF, Scott DA, Sanchez FG, Voldseth RA, Page-Dumroese D, Elioff JD, Stone DM: The North American long-term soil productivity experiment: Findings from the first decade of research. *For Ecol Manag* 2005, 220:31-50.
41. Puhlick JJ, Laughlin DC, Moore MM: Factors influencing Ponderosa Pine regeneration in the Southwestern USA. *For Ecol Manag* 2012, 264:10-19.
42. Raffa KF, Aukema BH, Bentz BJ, Carroll AL, Hicke JA, Turner MG, Romme WH: Cross-scale drivers of natural disturbances prone to anthropogenic amplification: the dynamics of bark beetle eruptions. *Bioscience* 2008, 58:501-517.
43. Rouault G, Candau J-N, Lieutier F, Nageleisen L-M, Martin J-C, Warzée N: Effects of drought and heat on forest insect populations in relation to the 2003 drought in western Europe. *Ann Forest Sci* 2006, 63:613-624.
44. Runyon J, Waring RH, Goward SN, Welles JM: Environmental limits on net primary production and light-use efficiency across the Oregon transect. *Ecol Appl* 1994, 4:226-237.
45. Samuelson LJ, Johnsen K, Stokes T: Production, allocation and stemwood growth efficiency of*Pinus taeda* L. stands in response to 6 years of intensive management. *For Ecol Manag* 2004, 192(1):59-70.

46. Schmid JM, Mata SA: *Stand density and mountain pine beetle-cause tree mortality in ponderosa pine stands. RN-RM-515. U.S.* Department of Agriculture, Forest Service, Rocky Mountain Forest and Range Experiment Station, For Collins, CO; 1992.
47. Schmid JM, Mata SA: *Mountain pine beetle-caused tree mortality in partially cut plots surrounded by unmanaged stands. RP-RMRS-54. U.S.* Department of Agriculture, Forest Service, Rocky Mountain Research Station, Fort Collins, CO; 2005.
48. Scianna J: *Ponderosa pine. USDA NRCS Plant Materials Technical Note, MT-73*. 2011.
49. Sperry JS, Hacke UG, Oren R, Comstock JP: Water deficits and hydraulic limits to leaf water supply. *Plant Cell Environ* 2002, 25:251-263.
50. Turner J, Lambert M: Soil and nutrient processes related to eucalypt forest dieback. *Austral for* 2005, 68:251-256.
51. Turner DP, Cohen WB, Kennedy RE, Fassnacht KS, Briggs JM: Relationships between Leaf Area Index and Landsat TM Spectral Vegetation Indices across Three Temperate Zone Sites. *Remote Sens Environ* 1999, 70(1):52-68.
52. USDA NRCS, Soil Survey Division Staff: *Soil Survey Manual. Soil Conservation Service. U.S. Department of Agriculture Handbook 18*. 1993.
53. USDA NRCS, Soil Survey Division Staff: *Soil Survey Geographic (SSURGO) Data Base: Data Use Information. US Department of Agriculture Miscellaneous Publication Number 1527*. 1995.
54. USFS FHTET: *USDA Forest Service, Forest Health Protection, Forest Health Technology Enterprise Team, data*. 2010.
55. USGS: *USGS Global Visualization Viewer*. 2014.
56. Viereck LA, Dyrness CT, Van Cleve K, and Foote MJ: Vegetation, soils, and forest productivity in selected forest types in interior Alaska. *Can J For Res* 1983, 13:703-720.
57. Villalba R: Tree-ring and glacial evidence for the medieval warm epoch and the little ice in southern South America. *Clim Change* 1990, 26:183-197.
58. Wallin KF, Raffa KF: Influences of external chemical cues and internal physiological parameters on the multiple steps of post-

landing host election behavior of *Ipspini*(Coleoptera: Scholytidae). *Environ Entomol* 2000, 29:442-453.

59. Wallin KF, Raffa KF: Feedback between individual host selection behavior and population dynamics in an eruptive insect herbivore. *Ecol Monogr* 2004, 74:101-116.
60. Waring RH: Estimating forest growth and efficiency in relation to canopy leaf area. *Adv Ecol Res* 1983, 13:327-354.
61. Waring RH, Pitman GB: Physiological stress in lodgepole pine as a precursor for mountain pine beetle attack. *J Appl Entomol* 1983, 1:265-270.
62. Waring RH, Coops NC, and Running SW: Predicting satellite-derived patterns of large-scale disturbances in forests of the Pacific Northwest region in response to recent climatic variation. *Remote Sens Environ* 2011, 115:3554-3566.
63. Waring RH, Coops NC, Mathys A, Hilker T, Latta G: Process-based modeling to assess the effects of recent climatic variation on site productivity and forest function across Western North America. *Forests* 2014, 5:518-534.
64. Wright LC, Berryman AA, Gurusiddaiah S: Host resistance to the fir engraver beetle,*Solytusventralis* (Coleoptera: scolytidae): 4. Effect of defoliation on wound monotermpene and inner bark carbohydrate concentrations. *Can Entomol* 1979, 111:1255-1262.
65. Wright LC, Berryman AA, Wickman BE: Abundance of the fir engraver, *Scolytusventralis*, and the Douglas-fir beetle, *Dentroctonus pseudostugae*, following tree defoliation by the Douglas-fir tussock moth, *Orgyia pseudstugata. Can Entomol* 1984, 116:293-305.
66. Wu C, Chen JM: Diverse responses of vegetation production to interannual summer drought in North America. *Int J Appl Earth Obs* 2013, 21:1-6.
67. Xu XN, Wang Q, Shibata H: Forest structure, productivity and soil properties in a subtropical evergreen broad-leaved forest in Okinawa, Japan. *J Forest Res* 2008, 19:271-276.
68. Yoshiko A, Mueller-Dombois D: A landscape perspective of the Hawaiian rain forest dieback. *J Veg Sci* 1995, 6:449-464.

69. Zausen GL, Kolb TE, Bailey JD, Wagner MR: Long-term impacts of stand management on ponderosa pine physiology and bark beetle abundance in northern Arizona: A replicated landscape study. *For Ecol Manag* 2005, 218:291-305.

70. Zhao J, Liu JL, Yang L: A preliminary study on mechanisms of LAI inversion saturation. *Int Arc Photogram Remote Sens Spatial Inf Sci* 2010, XXXIX-B1:77-81.

Chapter 5

Spatial Heterogeneity of Urban Soils: The Case of the Beijing Metropolitan Region, China

Qizheng Mao[1], Ganlin Huang[1], Alexander Buyantuev[2], Jianguo Wu[1, 3], Shanghua Luo[4], and Keming Ma[5]

[1]Center for Human-Environment System Sustainability (CHESS), State Key Laboratory of Earth Surface Processes and Resource Ecology (ESPRE), Beijing Normal University, Beijing 100875, China

[2]Geography and Planning, University at Albany, State University of New York, Albany 12222, NY, USA

[3]School of Life Sciences and Global Institute of Sustainability, Arizona State University, Tempe 85287, AZ, USA

[4]Chinese Research Academy of Environmental Sciences, Beijing 100012, China

[5]State Key Laboratory of Urban & Regional Ecology, Research Center for Eco-Environmental Sciences, Chinese Academy of Sciences, Beijing 100085, P. R. China

ABSTRACT

Introduction

Urban soils are large pools of carbon, nitrogen, and other elements, supporting plant growth, sustaining biogeochemical cycles, and serving as the foundation for maintaining ecosystem function and services of urban green spaces (UGS). Quantifying urban soil properties is essential for assessing urban ecosystem services and detecting pollution. Characterizing spatial heterogeneity of urban soil properties, which may change with land use or urbanization, is crucial for understanding urban ecosystem functions.

Methods

We collected 466 soil composite samples across the Beijing metropolitan region and then analyzed the heterogeneity of soil properties in UGS of Beijing metropolitan region by ANOVA along the urbanization gradient spanning different land uses.

Results

Our results show that soil properties vary considerably among land-use types and along the urban-rural gradient. Soil moisture content, soil organic carbon (SOC), available phosphorus, available potassium, total nitrogen (TN), the ratio of SOC:TN (C/N), Pb, and Cu increased notably from suburbs to the urban core, while soil pH, bulk density, TN, Cr, Ni, and Mn were not significantly different across urbanization levels. Most soil properties, except soil TN, Cu, Cr, Ni, exhibited significant differences between different lands uses. The highest levels of soil nutrients and heavy metals were found in roadsides and residential areas among all the land uses.

Conclusions

The spatial heterogeneity of urban soils in the Beijing metropolitan region is mainly attributable to the different land uses. Soil nutrient content has increased in UGS due to fertilization and pollution. Contamination of urban soils by heavy metals is due largely to traffic emissions and pollution by industrial and household wastes. Although human activities tend to augment urban soil nutrient pools to maximize certain ecosystem services, they may inadvertently compromise these and other services by increasing pollution. Knowledge of urban soil spatial heterogeneity in UGS is indispensable for improving urban ecosystem management.

INTRODUCTION

Soil, as the component of urban green spaces (UGS) (Setälä et al. [2013]), plays a pivotal role in maintaining urban ecosystem services, such as biodiversity maintenance, water resources protection, microclimate regulation, carbon sequestration, food production, and fulfillment of cultural or recreational needs (Lovell and Taylor [2013]; Neil et al. [2014]; Wu [2014]). Urban soils are large pools of carbon, nitrogen, and other elements, serving as the foundation of urban ecosystems by supporting plant growth and sustaining biogeochemical cycles (Pouyat et al. [2002]). Quantifying soil properties and understanding soil conditions are essential for assessing ecosystem services provided by UGS and detecting pollution in UGS.

Urban soils are often physically disturbed, chemically polluted, and tend to be highly heterogeneous spatially (Wu [2014]). Previous studies conducted at regional scales examined soil properties' variations along urbanization gradients of Hong Kong (Jim [1998a]), Phoenix (Zhu et al. [2006]), Baltimore, and New York (Pouyat et al. [2008]). Major findings from these and other studies are as follows. First, urbanization is accompanied by the transporting of large amounts of rock material in the landscape, the introduction of construction rubble, bricks, and other building materials (Jin et al. [2011]). Consequently, urban soils often are compacted and structurally degraded, alkaline, and water deficient (Jim [1998a]), which makes them less suitable for growing plants. Second, existing soil patterns are the product of both human disturbances and management-induced amelioration and therefore the effects of urbanization on soil nutrients. For example, some researcher demonstrated that urban soil had lower nutrients (Jim [1998b]; Baxter et al. [2002]), while other studies pointed out that soil nutrients were higher in urban area (Kaye et al. [2008]) and had sufficient nutrient levels to support plant growth (Pouyat et al. [2007]). Third, pollutants from industrial activities (Shang et al. [2012]), heavy traffic (Chen et al. [2010a]; Hamzeh et al. [2011]), and house wastes (Schwarz et al. [2012]; Szolnoki et al. [2013]) all contribute to soil contamination by heavy metals, which is likely to have adverse effects on human health.

At local scales, urban soil properties were compared between different land uses in a number of cities, including Baltimore (Pouyat et al. [2007]), Hong Kong (Jim [1998a]), Hubei (Li et al. [2013]), and Lanzhou (Wang et al. [2011]). It is widely agreed that the difference in human activities contribute to the variation of soil properties among land-use types (Pouyat et al. [2007]; Davies and Hall [2010]). For example, soils in transportation and residential areas were characterized by higher pH and bulk density (BD) (Pouyat et al. [2007]; Zhao et al. [2013]; Yang et al. [2014]). Soils in urban forests (Zhao et al. [2013]) or protected places (Yang et al. [2014]) contained higher soil organic carbon (SOC), while soils at industrial sites and in transportation areas were often polluted by heavy metals (Lu and Bai [2010]; Hamzeh et al. [2011]). At finer scales, urban soil carbon and nitrogen also varied within single patches, although this variation was smaller than that at broader scales and different land uses (Jenerette et al. [2006]).

Previous investigations undoubtedly suggest significant transformations in urban soil properties as a result of land use and land cover changes. However, a deep understanding of the urban soil biogeochemistry is still limited, requiring more research to be conducted. Considerable spatial heterogeneity of urban soils warrants a landscape ecological perspective for analyzing patterns and determinants of soils in urban areas. We therefore investigated urban soil variation in UGS of Beijing along the urbanization gradient and across different land-use types in order to understand this heterogeneity. Specifically, we focused on three questions: (1) what are the soil properties in UGS in Beijing and do they meet the needs of plant growth or not? (2) How and do soil properties vary along urbanization gradients or not? (3) How and do soil properties vary across land-use types or not? Our findings are expected to have implications for a better understanding of urbanization effects on soils and contribute to scientifically informed soil management and conservation in UGS.

METHODS

Study Area

Beijing, one of the oldest cities in the world which has experienced rapid urbanization in the recent decades, was chosen for our case study. It is located in the northeast of the North China Plain and has an area of approximately 16,410 km^2 and a population of 20.69 million (Beijing Municipal Bureau of Statistics [2013]). During 1984 and 2008, the urbanized area in Beijing increased 3.4 times (Kuang [2012]), while permanent residents increased by 53.4% (Beijing Municipal Bureau of Statistics [2013]). Beijing has diverse land uses including residential (52.64%), wood (7.04%), institutional (6.83%), industrial, and commercial (4.62%), transportation (2.91%), and others (Kuang [2012]). Such degree of urbanization and variability of land-use types makes Beijing a good study site to address our research questions.

Soil Sampling

In July 2009, we set up eight transects from the urban core to suburban areas in eight directions (east, west, north, south, northeast, northwest, southeast, southwest) in order to capture urbanization gradients. Along these transects, 78 survey plots spaced at 3-km interval were selected. Another 52 plots at 6-km interval were later added to densify our sampling network and capture more of the spatial heterogeneity. Consequently, 130 plots were sampled (Figure 1). We recorded dominant land use within the 500-m radius of each plot. One composite surface (0 to 20 cm) soil core was obtained within a representative 20 × 20-m open space by mixing three cater-cornered subsamples. Three additional samples were collected by cutting rings and then transferring them into three aluminum boxes. A total of 466 soil composite samples were collected during July to September.

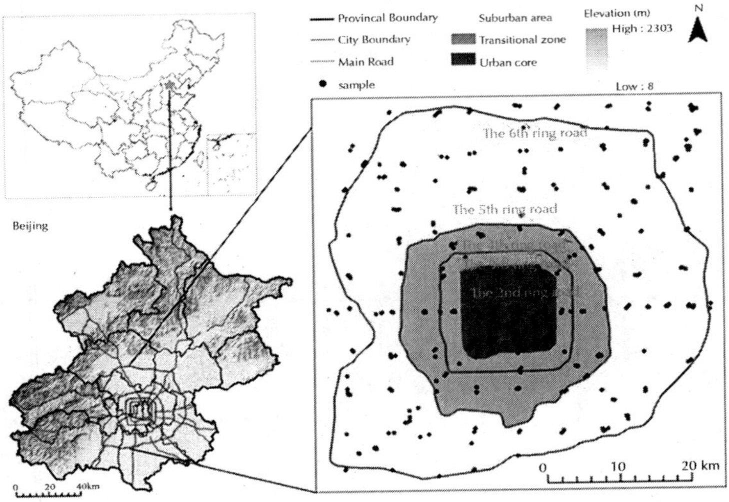

Figure 1: Study area and soil sampling design.

Sample Preparation and Instrument Analysis

Soil samples were stored indoor and dried at room temperature. Debris (e.g., leaves, stones, worms) were handpicked and removed. The samples were then grounded to go through 2-, 1-, and 0.15-mm nylon sieves, respectively. Laboratory analyses for soil properties were then conducted using soil composites through different nylon sieves. Specifically, we analyzed the soil moisture content (SMC), soil pH, soil BD, total nitrogen (TN) content, SOC:TN (C/N), SOC, available phosphorus (AP), available potassium (AK), and heavy metals (Pb, Cr, Cu, Ni, Mn).

All laboratory tests followed standard methods. Using soils from the aluminum boxes, SMC was determined by drying them at 105°C for 24 h and was calculated as the percentage of water in oven-dry soil (%). BD was determined from the total oven-dry mass and was calculated as milligrams of oven-dry soil in the volume of 1 cm^3. Using soils through 2-mm nylon sieves, pH was measured with a 1:2.5 mixture (soil:water ratio) by the glass electrode method (McLean [1982]). Using soils through 1-mm nylon sieves, we measured AP by Mo-Sb colorimetry from a sample extracted with $NaHCO_3$ and non-phosphorus active carbon (Tiessen et al. [1993]), while AK was extracted with ammonium acetate and examined directly by inductively coupled plasma optical emission spectroscopy (ICP-OES) (Simard [1993]). Using soils through 0.15-mm nylon sieves, TN and TC were analyzed by the dry combustion method (Vario EL Element Analyzer) (Bremner and Mulvaney [1982]). SOC was measured by potassium dichromate titration method (Lu [2000]). Finally, we measured the contents of Pb, Cu, Cr, Ni, and Mn by extracting them using HNO_3, $HClO_4$, and HF according to USEPA method 3052 with a microwave laboratory unit (USEPA United States Environmental Protection Agency [1995]). Pb, Cr, Mn, Cu, and Ni contents were all determined using ICP-OES.

Urbanization Gradient and Classification of Land Use

Beijing has five circular express roads, which are named as the second to the sixth ring road. The rapid urban sprawl in Beijing followed a homocentric expansion along this ring-road system (Kuang [2012]). We therefore use these ring roads as benchmarks of urbanization phases in Beijing. According to this logic, we define the area within the third ring road as the urban core area, while the area between the third and the fifth ring roads as the urban transitional zone. The suburban area encompasses developments outside the fifth ring road (Figure 1).

We categorized all 466 samples to seven classes according to the dominant land use and land cover type (Figure 2), and they were as follows:

- Park ($n=57$), open space for recreational use
- Residential area ($n=78$), residential buildings
- Institutional area ($n=40$), school, hospital, and office places
- Road greenbelts ($n=114$), large areas of ribbon vegetation distributed along roadsides
- Roadside ($n=104$), single tree rows along major roads
- Industrial area ($n=17$), factories
- Forest ($n=57$), large area of vegetation and plants in suburban areas

Figure 2: Seven types of land use in our study. (a) Road greenbelts,(b) roadside, (c) residential area, (d) neighborhood park, (e) forest, (f)industrial area, (g) institutional sites.

Statistical Analyses

Statistical analyses were performed in SPSS 21. The Kolmogorov-Smirnov (K-S) test for normality was used to determine whether variables were normal or skewed. Data were log transformed to stabilize the variance of soil properties for AP, AK, SOC, TN, C/N, Pb, Cu, Ni, and Mn. The analysis of variance (ANOVA) was used to test whether soil properties varied between different areas followed by multiple comparisons to test the effect of land-use types and urbanization on each variable. To deal with sample size inequalities for different land-use types or urbanization levels, we employed the Hochberg's method for *post hoc* analysis to determine significant differences between mean soil variables (Pouyat et al. [2007]).

Background Values of Soil Properties in Beijing

To articulate human activities' impacts on UGS soil properties, we compared our samples with background values of soil properties in Beijing (Su et al. [2001]; Liu et al. [2003]), which were sampled in mountains in the outskirt of Beijing to represent soil properties values without human disturbance. We refer to China's Environmental Protection Agency's work for these background soil values (China National Environmental Monitoring Center [1990]).

RESULTS

General characteristics of urban soils in Beijing's UGS

Descriptive statistics of each soil variable revealed that soil properties in UGS varied widely (Table 1). The highest coefficient of variation (CV) was found in soil AP (119.25%) and the lowest in soil pH (2.36%).

Table 1: Descriptive statistics of soil properties in Beijing

Soil properties	Median	Minimum	Maximum	SE	CV (%)	Mean	Background value	K-S*P* value
pH	8.07	7.43	8.77	0.01	2.36	8.05	6.47[a]	0.010
BD (mg/kg)	1.39	0.89	1.74	0.01	9.13	1.38	1.11[b]	0.054
Moisture (%)	17.35	3.79	54.84	0.28	34.90	17.02	16.62[a]	0.278
AP (mg/kg)	14.33	0.61	274.03	1.16	119.25	20.71	11.37[a]	<0.001
AK (mg/kg)	158.13	41.20	711.10	4.16	50.54	176.44	281.8[a]	<0.001
SOC (g/kg)	9.61	1.68	64.28	0.26	53.94	10.51	62.16[a]	<0.001
TN (g/kg)	0.85	0.23	2.46	0.01	33.91	0.89	2.84[a]	0.009
C/N	11.16	4.08	26.13	0.15	27.77	11.59	21.54[a]	<0.001
Pb (mg/kg)	18.53	0.66	291.98	0.83	81.45	21.80	25.40[c]	<0.001
Cr (mg/kg)	54.11	5.24	154.75	0.54	21.08	54.14	68.10[c]	<0.001
Cu (mg/kg)	25.15	0.70	125.22	0.64	47.65	28.39	23.60[c]	<0.001
Ni (mg/kg)	25.03	11.84	61.56	0.21	18.02	25.33	29.00[c]	0.010
Mn (mg/kg)	672.10	338.21	1572.52	5.76	17.84	685.49	705.00[c]	0.049

[a]Liu et al. [2003], a study of soil properties in mountain region, Beijing; [b]Su et al. [2001], another study of soil properties in mountain region, Beijing; [c]China National Environmental Monitoring Center [1990], a book of soil background value in China. SE-Standard errors; CV-Coefficient of variation.

Mao *et al.*

Mao *et al. Ecological Processes* 2014 3:23, doi: 10.1186/s13717-014-0023-8

Comparing our results with soil background values, we found that the mean values of soil pH, BD, and AP were higher, AK, SOC, TN, and C/N were lower, and SMC was similar to that in Dangling Mountain in Beijing. Of these, only mean Cu was higher than the soil background value of Beijing (23.6 mg/kg), which signifies relatively clean soils, close to natural soils.

Variation of Soil Properties along Urbanization Gradients

Table 2 shows variation of soil properties along the urbanization gradient. Soil moisture, AP, AK, SOC, C/N, Pb, and Cu showed significant increase from suburban area to the urban core (Figure 3). The SMC in suburban areas (16.25%) is not statistically different ($P=0.250$) from the transitional zone (17.4%) but both are significantly different from the urban core (20.12%). Soil AP and AK were significantly different between different urbanization levels ($P<0.001$), but the difference of SOC was only detected between the suburban area and the urban core ($P=0.012$), while the difference of soil C/N was detected between suburban area and other two areas. The soil Pb was significantly higher ($P<0.001$) in the urban core and the transitional zone compared to suburban areas, whereas no significant difference of soil Pb is detected between the urban core and the transitional zone ($P=0.659$). Soil Cu content was significantly higher in the urban core area when compared to the transitional zone and suburban areas. For soil pH, BD, TN, Cr, Ni and Mn, no significant difference was found along the urbanization gradient.

Table 2: Mean (± SE) surface soil properties (0 to 20 cm) along the urbanization gradient

Soil properties	Suburban area (*n* =305)	Transitional zone (*n* =98)	Urban core (*n* =63)	Background value	*P* value
pH	8.03 ± 0.01	8.07 ± 0.02	8.09 ± 0.02	6.47	0.076
BD (mg/m^3)	1.38 ± 0.01	1.39 ± 0.01	1.39 ± 0.02	1.11	0.836
SMC (%)	16.25 ± 0.33^a	17.4 ± 0.6^a	20.12 ± 0.79^b	16.62	<0.001
AP (mg/kg)	16.08 ± 1.31^a	22.09 ± 2^b	40.25 ± 3.9^c	11.37	<0.001
AK (mg/kg)	153.61 ± 4.21^a	188.83 ± 7.61^b	266.65 ± 14.66^c	281.8	<0.001
SOC (g/kg)	10.19 ± 0.35^a	10.76 ± 0.52^{ab}	11.68 ± 0.54^b	62.16	0.012
TN (mg/kg)	0.89 ± 0.01	0.86 ± 0.03	0.97 ± 0.05	2.84	0.09
C/N	11.24 ± 0.2^a	12.16 ± 0.3^b	12.33 ± 0.34^b	21.54	0.001
Pb (mg/kg)	18.84 ± 1.05^a	26.63 ± 1.57^b	29.56 ± 2.05^b	25.4	<0.001
Cr (mg/kg)	53.51 ± 0.68	56.37 ± 1.34	53.85 ± 0.69	68.1	0.074
Cu (mg/kg)	25.78 ± 0.59^a	28.98 ± 1.59^a	39.95 ± 2.32^b	18.7	<0.001
Ni (mg/kg)	25.25 ± 0.26	25.7 ± 0.59	25.2 ± 0.32	29	0.77
Mn (mg/kg)	686.58 ± 7.87	674.28 ± 9.98	697.09 ± 10.86	705	0.43

Values with the same letters in superscript are not significantly different at 5% level based on Hochberg's multiple comparison tests. Last column shows *P* values for ANOVA tests.

Mao *et al.*

Mao *et al. Ecological Processes* 2014 3:23, doi: 10.1186/s13717-014-0023-8

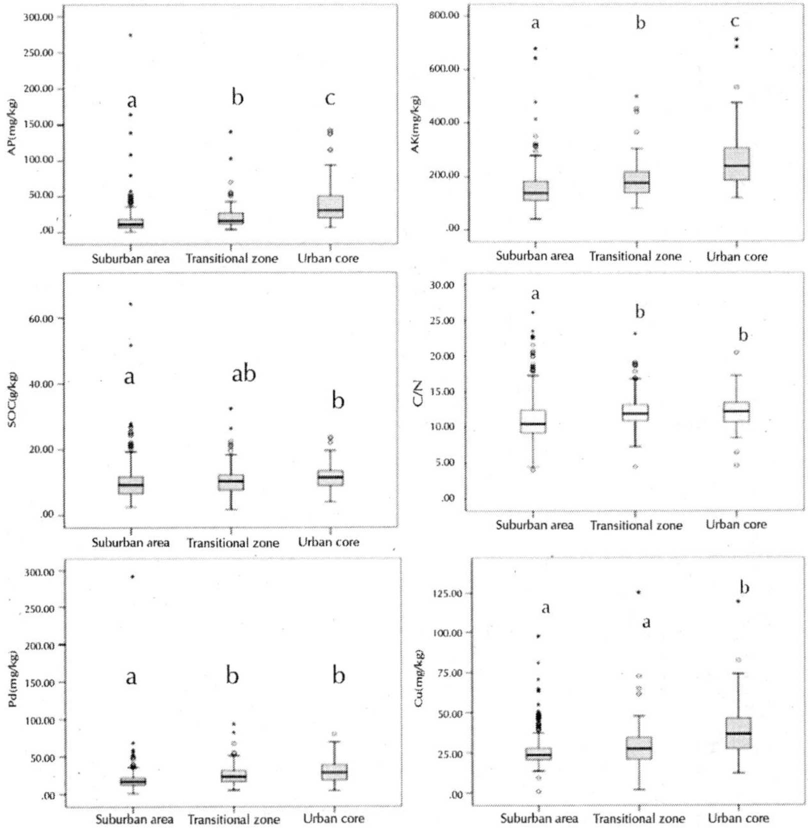

Figure 3: Variation of soil AP, AK, SOC, C/N, Pb, and Cu along the urbanization gradient. The different letters represent the significant difference between two land uses.

Variation of Soil Properties within Different Land-Use Types

The soil properties were notably different between land-use types (Table 3). Our results showed that soil pH, BD, SMC, AP, AK, SOC, C/N, TN, Pb, and Mn were significantly different between the seven land-use types ($P<0.05$). Detailed results on how each soil property varies across land-use types are listed in Table 3. In particular, we want to point out the following soil property variations across land-use types and will discuss what caused such variation in the 'Discussion' section.

Table 3: Mean (± SE) surface soil properties (0 to 20 cm) of different land uses

Soil properties	**Industrial**	**Roadside**	**Residential**	**Institutional**	**Road greenbelts**	**Park**	**Forest**	***P* value**
pH	8.02 ± 0.04 ab	8.12 ± 0.02 a	8.08 ± 0.02 a	8.10 ± 0.02 a	8.05 ± 0.02 a	7.96 ± 0.03 b	7.95 ± 0.02 b	<0.001
BD (mg/m3)	1.32 ± 0.04 ab	1.43 ± 0.01 a	1.42 ± 0.01 a	1.41 ± 0.02 a	1.38 ± 0.01 a	1.38 ± 0.01 a	1.28 ± 0.02 b	<0.001
SMC (%)	19.09 ± 0.84 ac	13.01 ± 0.66 b	19.50 ± 0.46 a	18.71 ± 0.67 ac	18.11 ± 0.57 ac	17.77 ± 0.62 ac	16.25 ± 0.74 c	<0.001
AP (mg/kg)	31.15 ± 17.54 ab	19.33 ± 2.30 ab	23.34 ± 2.04 a	18.48 ± 2.58 ab	21.79 ± 1.87 a	20.52 ± 3.9 ab	16.15 ± 3.28 b	0.003
AK (mg/kg)	145.36 ± 16.85 ab	209.76 ± 9.98 a	206.93 ± 12.42 a	149.84 ± 9.93 b	163.49 ± 6.37 b	137.17 ± 7.1 b	166.14 ± 12.48 b	<0.001
SOC (g/kg)	15.04 ± 4.18 ab	13.09 ± 0.64 a	10.22 ± 0.43 b	8.52 ± 0.49 b	9.46 ± 0.44 b	9.59 ± 0.64 b	9.57 ± 0.49 b	<0.001
TN (g/kg)	1.07 ± 0.15	0.93 ± 0.03	0.88 ± 0.02	0.81 ± 0.03	0.84 ± 0.03	0.91 ± 0.05	0.93 ± 0.04	0.052
C/N	12.02 ± 1.06 ab	13.93 ± 0.37 b	11.39 ± 0.3 a	10.71 ± 0.47 a	11 ± 0.25 a	10.47 ± 0.33 a	10.24 ± 0.3 a	<0.001
Pb (mg/kg)	41.51 ± 18.25 ab	24.74 ± 1.51 a	23.90 ± 1.70 ab	20.46 ± 1.89 ab	19.11 ± 0.97 b	16.58 ± 0.9 b	19.79 ± 1.17 ab	0.010
Cr (mg/kg)	54.03 ± 2.36	55.56 ± 1.42	55.11 ± 1.07	51.99 ± 2.16	52.64 ± 1.04	53.69 ± 1.54	55.19 ± 0.95	0.279
Cu (mg/kg)	28.82 ± 4.28	31.63 ± 1.66	30.17 ± 1.54	25.71 ± 1.47	26.76 ± 1.09	27.39 ± 2.10	26.12 ± 1.25	0.068
Ni (mg/kg)	25.40 ± 1.33	24.35 ± 0.53	26.13 ± 0.43	25.30 ± 0.61	25.15 ± 0.30	25.61 ± 0.84	26.11 ± 0.63	0.071
Mn (mg/kg)	708.08 ± 73.56 ab	721.64 ± 11.83 a	683.78 ± 11.33 ab	677.10 ± 17.58 ab	673.68 ± 9.19 ab	681.71 ± 18.16 ab	646.30 ± 17.45 b	0.004

Values with same letters are not significantly different at 5% level based on Hochberg's multiple comparison tests. Last column shows *P* values for ANOVA tests.

Mao *et al.*

Mao *et al. Ecological Processes* 2014 3:23, doi: 10.1186/s13717-014-0023-8

The soil pH, BD, and AP in roadsides and residential areas are significantly higher than in the forests. The soil Mn, SOC, and C/N are higher in roadsides than the rest land-use types. Soil AP is significantly higher in road greenbelts.

DISCUSSION

Soil Conditions of UGS in Beijing

Plant Growth

Our results indicate that soils are practically unsuitable for plant growth in many places within the Beijing metropolitan region. The soils of UGS in this region are commonly alkaline and mechanically compacted. The soil pH in all sites exceeds the pH level of neutral soils (pH=7), which makes them unsuitable for growing most of plants according to earlier recommendations (Whitcomb [1987]). The high levels of alkaline can influence soil nutrient solubility (Messenger [1986]) and lead to soil fertility imbalance, deficiency, and chlorosis symptoms (Harrell et al. [1988]). Furthermore, measured soil BD was above 1.10 mg/m^3 for 98.92% of the samples and 1.49 mg/m^3 for 18.2% of the samples, which are expected to affect plant root growth (USDA United States Department of Agriculture [2001]). This is also aggravated by reduced soil moisture holding capacity, aeration, and drainage.

Our results also suggest deficiencies in soil AP and SOC. While soil AP in urban sites is almost twice as high (11.67 mg/kg) as the background value (Liu et al. [2003]), it is still lower (less than 30 mg/kg in 82.17% of the samples) than the recommended suitability range of 30 to 60 mg/kg for healthy plant growth (Whitcomb [1987]). The SOC in the urban area (10.51 g/kg) is more than five times less than the background value (62.16 g/kg) (Liu et al. [2003]). Moreover, only 42.25% of the samples are above the recommended range of soil SOC (>10 g/kg) for healthy plant growing (Whitcomb [1987]). Such SOC deficiency in UGS can easily deplete soil structural composition and complicate maintenance (Jim [1998b]).

In contrast to soil AP and SOC, AK is found in sufficient, even excessive, quantities in Beijing (176.44 mg/kg), which is higher than the recommended range of 78 to 156 mg/kg for healthy plant growth (Whitcomb [1987]). The negative side of very high AK in about 50% of the samples is the potential imbalance in nutrients, which has consequences for Ca absorption (Hamberg et al. [2009]; Moussa et al. [2009]).

Pollution

The overall health risk of heavy metal pollution in soils of green spaces in Beijing is relatively low whereas some higher localized soil Pb and Cu content exists. Cu in 18.54% of the samples, Pb in 11.23% of the samples, and Cr and Ni in 0.66% of the samples exceed the environmental quality standards in China (35 mg/kg for Cu and Pb, 90 mg/kg for Cr, and 40 mg/kg for Ni). Two out of the 466 samples of Cu and one sample of Ni exceed the second level (100 mg/kg for Cu and 60 mg/kg for Ni). This level corresponds to the safety threshold for human health (SEPA China State Enviromental Protection Administration of China [1995]).

Effects of Urbanization on Soil Properties

Our findings indicate that soil properties change considerably along urbanization gradients, which is consistent with previous researches (Jim [1998a]; Zhu et al. [2006]; Kaye et al. [2008]; Pouyat et al. [2008]) (Figure 3). Human disturbance impacts soil properties in UGS mostly through intensive management and waste discharge.

The urban core has significantly higher SMC than the transitional zone and suburban area because UGS in the urban core are intensively maintained and frequently irrigated. Our finding of the increased SMC in the urban core of Beijing echoed the patterns of Phoenix in Arizona, USA (Zhu et al. [2006]).

Frequent irrigation and fertilization in urban areas also caused dramatic increase in soil AP, AK, and SOC from suburbs to the urban core (Figure 3), which is consistent with the higher content of soil nutrients in urban soils found previously (Zhu et al. [2006]; Kaye et al. [2008]; Pouyat et al.[2008], [2010]). Direct inputs of pollutants also contribute to such patterns (Kaye et al. [2008]). For example, untreated sewage, domestic waste, and excessive fertilizer applications could all lead to AP accumulation in urban and suburban soils (Huang et al. [2013]). The urban construction, which uses cement and concrete with abundant Ca, is a source of higher surface charge densities and cation exchange interaction. Its most direct consequence is the fixation of absorbed K ions from soil colloids (Li et al. [2013]; Moussa et al. [2009]).

Our results show that SOC is positively correlated with the soil Pb ($R^2=0.41$, $P<0.001$) and Cu ($R^2=0.18$, $P<0.001$) (Figure 4). This finding is consistent with what researchers found previously that soil organic matter has a positive contribution to heavy metal accumulation (Sauvé et al. [2000]).

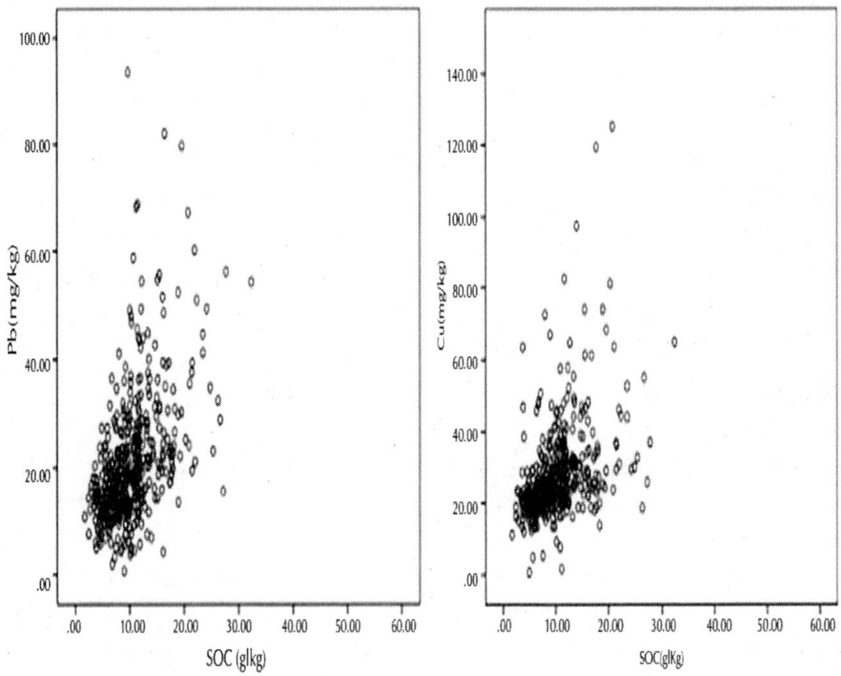

Figure 4: Correlation of soil Pb/Cu with SOC.

Soil pollution by heavy metals (i.e., Pb and Cu) increased significantly along the urbanization gradient (Table 2). The higher Pb concentrations in the urban core are explained by traffic emission and historical use of leaded paints (Chen et al. [2010b]; Xia et al. [2011]). Mao et al. ([2014]) presented an in-depth analysis on spatial heterogeneity of Pb in Beijing UGS.

Three factors contributed to the increase of soil Cu from the suburbs (25.78 mg/kg) to the urban core (39.95 mg/kg) (Figure 3). First, abundant historical sites (temples, palaces, ancient parks) in central Beijing contain many brass cultural relics. Second, Cu is known to be emitted as a result of deterioration of some mechanical parts in vehicles over time. In particular, the lining in automobile brakes was found to the source of Cu (Hamzeh et al. [2011]). Third, application of copper-rich pesticides in UGS may result in an increase of soil Cu in the urban core (Szolnoki et al. [2013]).

Effects of Land Use on Soil Properties

Most soil properties varied among seven land-use types, and these variations are due mainly to human-induced disturbances, as well as management practices that aim to improve soil conditions. Our results suggest that soils in roadsides and residential areas were most mechanically disturbed and had higher soil nutrients and heavy metal pollutants when compared to other land uses. This finding is consistent with previous findings (Pouyat et al. [2007]; Raciti et al. [2008]; Davies and Hall [2010]; Li et al. [2013]).

Roadsides and Residential Areas

Close to both streets and pedestrians, roadsides and residential areas are exposed to intense human disturbance, such as trampling, sewage discharge, and pet litter (Li et al. [2013]; Yang et al. [2014]). In contrast, forests in our study are located in infrequently disturbed suburbs away from busy streets and sidewalks (Figure 2e). As a result, soil pH and BD in roadsides and residential areas are significantly higher than in forests.

Roadside and residential areas have higher soil AK ($P<0.05$). Researchers found that the high level of Ca will increase the level of extractive K and such process is catalyzed by high soil pH (Li et al. [2013]). Construction materials (i.e., bricks and cement) with high Ca content are a major component in roadside and residential areas.

Therefore, we speculate that the high K in roadside and residential areas is related to the interaction between K and Ca in an alkaline environment. Soil AP is significantly higher in residential areas, which is mainly due to reclaimed water irrigation in these land uses. Roadsides have the lowest SMC. We attribute this to three factors. First, roadsides are less irrigated than UGS in other land uses. Second, proximity to roads and heavy automobile traffic can increase the ambient temperature and soil evaporation (Figure 2b). Third, higher soil BD decreases soil permeability and water infiltration thus increasing runoff.

Roadsides have significantly higher SOC, which can be explained by the following. First, soils in roadsides are seldom disturbed, while other land uses undergo transformations and replacement of vegetation by structures or other vegetation, which depletes SOC. Second, atmospheric deposition from vehicles and high accumulation of dust from nearby sealed surfaces result in increased nutrient content in roadsides (Chon et al. [1998]; Park et al. [2010]; Li et al. [2013]). Third, street litter and animal waste in roadsides may also lead to SOC accumulation (Park et al. [2010]; Zhao et al. [2013]). Similar to SOC, the highest C/N is found in roadsides followed by industrial areas. However, soil TN shows no statistical difference ($P=0.052$) between seven land-use types.

Soil Mn is also higher in roadsides. Li et al. ([2013]) also found highest soil Mn concentrations in this land-use type. The pattern can be explained by higher soil temperature near paved surfaces, which promotes soil mineralization and results in higher nutrient content (Curtin et al. [2011]).

Road Greenbelts

Soil AP is significantly higher in road greenbelts. We found parts of road greenbelts being irrigated by a mixture of reclaimed water from various urban sewages rich in phosphorous. The highest AP content (274.03 mg/kg) was detected near the steel production factory, the Shougang Construction Group. However, soil Pb and SOC in road greenbelts are relatively low, which is consistent with previous findings (Zhao et al. [2013]). This is because most of the road greenbelts in Beijing were newly built and filled with garden soils and have only experienced a short-time accumulation of soil organic matters and pollutants.

Limitations

Due to limited time and resources, we only measured selected soil properties, while other essential indicators of soil health, such as available nitrogen, micronutrients (e.g., Ca, Mg, Fe), and other heavy metal pollutants (e.g., Cd, Zn, as), were not assessed.

CONCLUSIONS

Urban soils in the Beijing metropolitan region are considerably alkaline and compacted. Soil TN, SOC, and AP are in deficit, while AK is abundant and sufficient for supporting plant growing. Heavy metal pollution in Beijing is low. Analyses of the spatial patterns of soil properties in Beijing suggest that soil AP, AK, SOC, C/N, Pb, and Cu increase from suburbs to the urban core, while other elements showed no significant difference. Land use was found as a strong predictor of spatial heterogeneity of soil properties. Roadsides and residential areas are the two land uses characterized by higher soil nutrients and heavy metal pollutants. The major factor of soil spatial heterogeneity in our study area is the time of soil accumulation followed by human activities (e.g., physical disturbance, pollutant emissions, irrigation, fertilization, and soil restoration). Our study is an important step toward understanding the spatiotemporal pattern of soil properties in urban regions. Future research should systematically evaluate both the negative and positive effects of management practices on urban soils and help design effective management plans for ameliorating soil conditions. Also, it is critically important to enumerate the different ecosystem services (and disservices) provided by urban soils.

AUTHORS' CONTRIBUTIONS

QM and KM designed the research; QM and SL conducted the research and analyzed the data; QM drafted the manuscript; GH, AB, and JW revised the manuscript. All authors read and approved the final manuscript.

ACKNOWLEDGEMENTS

This research was supported by two projects - the 'Special Foundation of State Key Lab of Urban and Regional Ecology (grant SKLURE2013-1-01)' and the 'Foundation for The Recruitment Program of Global Experts of Beijing Normal University (257-107023)'. The authors sincerely thank Chunyang He, Deyong Yu, and Zexiang Sun for their valuable input.

REFERENCES

1. Baxter JW, Pickett ST, Dighton J, Carreiro MM (2002) Nitrogen and phosphorus availability in oak forest stands exposed to contrasting anthropogenic impacts. Soil Biology and Biochemistry 34:623-633
2. (2013) Beijing statistical yearbook.
3. Bremner J, Mulvaney C (1982) Nitrogen—total. In: Miller RH, Keeney DR (eds) Methods of soil analysis, part 2, chemical and microbiological properties, American Society of Agronomy, Madison. pp 595-624
4. Chen C, Wang EL, Yu Q (2010) Modeling wheat and maize productivity as affected by climate variation and irrigation supply in North China plain. Agron J 102:1037-1049
5. Chen X, Xia X, Zhao Y, Zhang P (2010) Heavy metal concentrations in roadside soils and correlation with urban traffic in Beijing, China. J Hazard Mater 181:640-646
6. (1990) the backgrounds value of soil elements in China. China Environment Science, Beijing.
7. Chon H-T, Ahn J-S, Jung MC (1998) Seasonal variations and chemical forms of heavy metals in soils and dusts from the satellite cities of Seoul, Korea. Environ Geochem Health 20:77-86
8. Curtin D, Beare MH, Chantigny MH, Greenfield LG (2011) Controls on the extractability of soil organic matter in water over the 20 to 80°C temperature range. Soil Sci Soc Am J 75:1423-1430
9. Davies R, Hall SJ (2010) Direct and indirect effects of urbanization on soil and plant nutrients in desert ecosystems of the Phoenix

metropolitan area, Arizona (USA). Urban Ecosyst 13:295-317

10. Hamberg L, Malmivaara-Lamsa M, Lehvavirta S, Kotze DJ (2009) the effects of soil fertility on the abundance of rowan (Sorbus aucuparia L.) in urban forests. Plant Ecol 204:21-32
11. Hamzeh MA, Aftabi A, Mirzaee M (2011) Assessing geochemical influence of traffic and other vehicle-related activities on heavy metal contamination in urban soils of Kerman city, using a GIS-based approach. Environ Geochem Health 33:577-594
12. Harrell MO, Pierce P, Mootr D (1988) Pin oak and silver maple chlorosis treatment with ferric ammonium citrate solution. J Arboric 14:156-158
13. Huang L, Wang H, Li Y, Lu S (2013) Spatial distribution and risk assessment of phosphorus loss potential in urban-suburban soil of Lishui, China. Catena 100:42-49
14. Jenerette GD, Wu J, Grimm NB, Hope D (2006) Points, patches, and regions: scaling soil biogeochemical patterns in an urbanized arid ecosystem. Glob Chang Biol 12:1532-1544
15. Jim CY (1998) Soil characteristics and management in an urban park in Hong Kong. Environ Manag 22:683-695
16. Jim CY (1998) urban soil characteristics and limitations for landscape planting in Hong Kong. Landsc Urban Plan 40:235-249
17. Jin J-W, Ye H-C, Xu Y-F, Shen C-Y, Huang Y-F (2011) Spatial and temporal patterns of soil fertility quality and analysis of related factors in urban-rural transition zone of Beijing. Afr J Biotechnol 10:10948-10956
18. Kaye J, Majumdar A, Gries C, Buyantuyev A, Grimm N, Hope D, Jenerette G, Zhu W, Baker L (2008) Hierarchical Bayesian scaling of soil properties across urban, agricultural, and desert ecosystems. Ecol Appl 18:132-145
19. Kuang W (2012) Spatio-temporal patterns of intra-urban land use change in Beijing, China between 1984 and 2008. Chin Geogr Sci 22:210-220
20. Li Z-G, Zhang G-S, Liu Y, Wan K-Y, Zhang R-H, Chen F (2013) Soil nutrient assessment for urban ecosystems in Hubei, China. Plos One 8:e75856

21. Liu S, Ma K, Fu BJ, Kang Y, Zhang J, Yuxin Z (2003) The relationship between land form, soil characteristic and plant community structure in the Donglingshan mountain region, Beijing. Acta Phytoecologica Sinica 27:496-502

22. Lovell ST, Taylor JR (2013) Supplying urban ecosystem services through multifunctional green infrastructure in the United States. Landsc Ecol 28:1447-146

23. Lu R (2000) Analytical methods of soil agrochemistry. China Agricultural Science and Technology, Beijing.

24. Lu SG, Bai SQ (2010) Contamination and potential mobility assessment of heavy metals in urban soils of Hangzhou, China: relationship with different land uses. Environmental Earth Sciences 60:1481-1490

25. Mao Q, Ganlin H, Keming M, Sun Z (2014) Variations of Soil Lead in Different Land Uses Along the Urbanization Gradient in the Beijing Metropolitan Area. Int. J. Environ. Res. Public Health 11:3199-3214

26. McLean E (1982) Soil pH and lime requirement. Methods of soil analysis. Part 2. American Society of Agronomy Madison, Wisconsin, USA.

27. Messenger S (1986) alkaline runoff, soil pH and white oak manganese deficiency. Tree Physiol 2:317-325

28. Moussa AS, van Rensburg L, Kellner K, Bationo A (2009) Exploring differences of soil quality as related to management in semiarid rangelands in the western Bophirima District, North West province, South Africa. African Journal of Range and Forage Science 26:27-36

29. Neil K, Wu J, Bang C, Faeth S (2014) Urbanization affects plant flowering phenology and pollinator community: effects of water availability and land cover. Ecol Process 3:17

30. Park S-J, Cheng Z, Yang H, Morris EE, Sutherland M, Gardener BBM, Grewal PS (2010) Differences in soil chemical properties with distance to roads and age of development in urban areas. Urban Ecosyst 13:483-497

31. Pouyat R, Groffman P, Yesilonis I, Hernandez L (2002) Soil carbon pools and fluxes in urban ecosystems. Environmental Pollution 116 Supplement 1:S107-S118

32. Pouyat RV, Yesilonis ID, Russell-Anelli J, Neerchal NK (2007) Soil chemical and physical properties that differentiate urban land-use and cover types. Soil Sci Soc Am J 71:1010-1019
33. Pouyat R, Yesilonis I, Szlavecz K, Csuzdi C, Hornung E, Korsós Z, Russell-Anelli J, Giorgio V (2008) Response of forest soil properties to urbanization gradients in three metropolitan areas. Landsc Ecol 23:1187-1203
34. Pouyat RV, Szlavecz K, Yesilonis ID, Groffman PM, Schwarz K (2010) Chemical, physical, and biological characteristics of urban soils. In: Aitkenhead-Peterson J, Volder A (eds) Urban ecosystem ecology, American Society of Agronomy, Crop Science Society of America, Soil Science Society of America, Madison. pp 119-152
35. Raciti SM, Groffman P, Fahey T (2008) Nitrogen retention in urban lawns and forests. Ecological Applications 18:1615-1626
36. Sauvé S, Hendershot W, Allen HE (2000) Solid-solution partitioning of metals in contaminated soils: dependence on pH, total metal burden, and organic matter. Environment Science and Technology 34:1135-1131
37. Schwarz K, Pickett STA, Lathrop RG, Weathers KC, Pouyat RV, Cadenasso ML (2012) the effects of the urban built environment on the spatial distribution of lead in residential soils. Environ Pollut 163:32-39
38. (1995) GB15618-1995, Soil Environmental Quality Standards in China (In Chinese).
39. Setälä H, Bardgett R, Birkhofer K, Brady M, Byrne L, de Ruiter P, de Vries F, Gardi C, Hedlund K, Hemerik L, Hotes S, Liiri M, Mortimer SR, Pavao-Zuckerman M, Pouyat R, Tsiafouli M, van der Putten WH (2013) Urban and agricultural soils: conflicts and trade-offs in the optimization of ecosystem services. Urban Ecosyst 17:239-253
40. Shang LY, Sun RH, Wang ZM, Ji YH, Chen LD (2012) Assessment of heavy metal pollution in surface sediments of rivers in northern area of Haihe River Basin, China. Huan Jing Ke Xue 33:606-611
41. Simard R (1993) Ammonium acetate-extractable elements. Soil sampling and methods of analysis. Lewis, Boca Raton.
42. Su B, Han X, Qu C, Huang J (2001) Net nitrogen mineralization and trification in one pure pine (*Pinus tabulaeformis*) forests and

one pine-oak mixed forests in Dongling mountain region. Acta Phytoecologica Sinica 5:203

43. Szolnoki Z, Farsang A, Puskás I (2013) Cumulative impacts of human activities on urban garden soils: origin and accumulation of metals. Environ Pollut 177:106-115
44. Tiessen H, Moir J, Carter M (1993) Characterization of available P by sequential extraction. Soil Sampling and Methods of Analysis 7:5-229
45. (2001) Soil quality test kit guide.
46. (1995) Method 3052: Microwave assisted acid digestion of siliceous and organically based matrices. Test methods for evaluating solid waste.
47. Wang B, Zhao S, Xia DS, Yu Y, Tian SL, Jia J, Jiang XR (2011) Characteristics of heavy metal elements and their relationship with magnetic properties of river sediment from urban area in Lanzhou. Huan Jing Ke Xue 32:1430-1440
48. Whitcomb CE (1987) Establishment and maintenance of landscape plants. Lacebark
49. Wu J (2014) urban ecology and sustainability: the state-of-the-science and future directions. Landsc Urban Plan 125:209-221
50. Xia X, Chen X, Liu R, Liu H (2011) Heavy metals in urban soils with various types of land use in Beijing, China. J Hazard Mater 186:2043-2050
51. Yang L, Li Y, Peng K, Wu S (2014) Nutrients and heavy metals in urban soils under different green space types in Anji, China. Catena 115:39-46
52. Zhao D, Li F, Yang Q, Wang R, Song Y, Tao Y (2013) the influence of different types of urban land use on soil microbial biomass and functional diversity in Beijing, China. Soil Use Manag 29:230-239
53. Zhu W-X, Hope D, Gries C, Grimm NB (2006) Soil characteristics and the accumulation of inorganic nitrogen in an arid urban ecosystem. Ecosystems 9:711-724

Chapter 6

The Effect of Increase in the Temperature on the for Aging of Acromyrmex lobicornis (Hymenoptera: Formicidae)

Rodrigo Tizón[1], Juan Pedro Wulff[2], and Daniel Valerio Peláez[3]

[1]CERZOS-CONICET, CCT-Bahía Blanca, San Andrés 850, Bahía Blanca 8000, Argentina

[2]Departamento de Biología, Bioq. Y Farm, UNS; CERZOS-CONICET, CCT-Bahía Blanca, San Andrés 850, Bahía Blanca 8000, Argentina

[3]CIC; CERZOS-CONICET, CCT-Bahía Blanca, San Andrés 850, Bahía Blanca 8000, Argentina

ABSTRACT

Background

The leaf-cutting ants appear to be a suitable group for studying the effect of global warming on ectothermic animals. These ants of the *Atta* and *Acromyrmex* genera are considered to be the main herbivores in the Neotropics. They have patterns of harvesting associated with ranges of temperature. *Acromyrmex lobicornis* has the widest geographical distribution in Argentina. Peak activity has always been recorded in spring and at the beginning of autumn, being slightly lower in summer when they forage at night, and activity ceases almost completely in winter. In order to evaluate how an increase in temperature affects the activity of *A. lobicornis*, we studied the amount of foraging and the trophic preferences in two treatments under controlled humidity and temperature conditions (4.5°C) and we also measured the walking speed of the workers as a function of an increase in temperature (6°C to 32°C).

Results

The rate of harvesting was 33% higher in the warmer treatment. There was a tendency for the ten forage items evaluated to be harvested at the higher temperature. The trophic preference, with or without heat, showed some variation for different items: the shoots of *Olea europaea* and the dry *Schinus molle* were the most harvested with heat and without heat, respectively.

Conclusions

Our results support the hypothesis that the ants' activity changes at higher temperature, with higher rates of harvesting and a change in walking speed is observed. There is also variation in the trophic preference, selecting plant items with a higher composition of elements that are degradable by symbiotic fungi. The average walking speed of the workers increased gradually as a function of an increase of temperature up to a maximum speed of 2.85 cm/s. Our results suggest that small variations in ambient temperature significantly affect certain behavior patterns in the leaf-cutting ants.

BACKGROUND

The climate change concept refers to a modification in the state of the climate that is identifiable by an alteration in the mean and/or in the variability of its properties that persists for a prolonged period of time, measured in hundreds of years (IPCC [2007]). Global warming is unequivocal, and a sequence of events over the last 100 years makes this evident: increase in mean global temperature of Earth's surface (+0.74°C), retreat of glaciers and permanent snow in high mountains, rise in the sea level, and increased frequency of events such as floods and droughts on a global scale (IPCC [2007]). Weather maps were made for Argentina according to the projections generated by the climate model developed by CIMA/CONICET (Nuñez et al. [2005]). Using this model, the present climate, represented by the 1981 to 1990 decade, was compared with the 2081 to 2090 decade to show the trend in climate change in southern South America over the next 100 years (Nuñez [2009]). The following greenhouse gas emission scenarios were highlighted: A2 and B2 (increasing emission of CO_2 intermediate between emission scenarios), A1F1 (more extreme), and B1 (with the least change) (IPCC [2001], [2007]). The emission scenarios A2 and B2 were selected to make the model, as they are intermediate between the scenarios proposed by IPCC ([2001], [2007]). Under scenario A2, the temperature shows a positive tendency for all geographic areas (although there is also seasonal variation); a marked N-S (north-south) gradient is seen with an increase in mean annual temperature of between 4°C and 4.5°C for the N of Argentina and between 2.5°C to 3°C and 2°C to 2.5°C for the center and S respectively, whereas for the Andes, there is a N-S gradient between 4.5° and 2°C. The extreme value was detected in scenario A2 in the spring with an increase of 6°C in the mean seasonal temperature in the N of Argentina. In scenario B2, the spatial distribution of warming is similar to scenario A2 but with values between 2.5°C to 3°C, 1.5°C to 2°C, and of 1°C for the N, center, and S of the country, respectively (Nuñez et al. [2005]; Nuñez [2009]).

Many studies have analyzed the biotic or abiotic factors that define the geographic gradients of the richness, diversity, and abundance of species, and in all cases, *a priori*, a positive correlation with the temperature is seen (e.g., Kaspari [2004]; Kaspari et al. [2000a], [b], [2004]; Sanders et al.[2007]; Buckley and Jetz [2008]; Storch [2010]). The survival and reproduction of an organism are related to its capacity to become adapted to the limitations imposed by the environment, sometimes thermic, that directly affect the level of biochemical reactions, and interact with the physiology and behavior of the organism (Angilletta et al. [2002]; Hochochka and Somero [2002]). The magnitude of which an environmental factor acts as a limit for the survival and reproduction of an organism is a function that varies according to the sensibility of the individual to the environmental factor, and in particular, the ectotherms are more closely related to variations in ambient temperature than organisms capable of regulating their body temperature through metabolism (Huey and Stevenson [1979]; Angilletta et al. [2002]; Cooper et al. [2008]). Thermoregulation is not exclusive to endotherms (May [1979]); however, in general, for ectotherms there is an optimum temperature range and a narrow range of tolerance in which they are capable of carrying out their physiological functions, such as locomotion, growth, and reproduction (Huey and Stevenson [1979]; Angilletta et al. [2002]; Cooper et al. [2008]; Deutsch et al. [2008]). Insects comprise the majority of terrestrial biodiversity (Bale et al. [2002]), and as ectotherms they are vulnerable to variation in temperature (Angilletta et al. [2002]; Deutsch et al. [2008]). A common approach, seen in work on insects of the effects of temperature on their behavioral habits and physiological processes, is the study of functions such as locomotion (flight) and reproduction or growth rate (May [1979]; Bale et al. [2002]). Habits that are linked to variation in temperature are the way they search for food and other forms of locomotion, e.g., walking over the ground (Vogt et al. [2003]; Angilletta et al. [2008]; Jumbam et al. [2008]). Ants are the group most often used to study these interactions as they lack anatomical structures for flight, with the exception of the initial sexual state. Moreover, it is a tax on that shows variable feeding habits and different types of behavior, such as hunting, collecting (seeds and other dead animals), harvesting leaves, and cultivation of symbiotic fungi (Hölldobler and Wilson [1990]).

In general, ants forage (search and collect) when the ambient temperature is between 10°C and 40°C (Hölldobler and Wilson [1990]). Temperatures outside this range may be stressful, as they are close to the limits of thermic tolerance and survival (Bestelmeyer [2000]; Hartley and Lester [2003]; Angilletta et al. [2007]; Jumbam et al. [2008]; Bollazzi et al. [2008]; Bollazzi and Roces [2011]). However, some species live in deserts, e.g., *Pogonomyrmex* spp. that forage with soil temperatures higher than 50°C (Pol and Lopez de Casenave [2004]; Bucy and Breed [2006]; Cole et al. [2010]). The soil temperature is directly related to the ambient temperature, especially in semiarid regions with a large percentage of bare ground (Hölldobler and Wilson [1990]; Vogt et al. [2003]). The extreme temperatures are considered to be the most significant stress factor for ants in arid and semiarid regions. Temperature directly affects foraging activities due to its effects on oxygen consumption, water loss, and transport cost. However, several species of ants have become adapted to high temperatures and they show various strategies for enduring or avoiding these situations (Pol and Lopez de Casenave [2004]; Cole et al. [2010]). Temperature is a key factor for different groups of ants that compete for resources at the same site, to such an extent that they become competitive, and there is a positive correlation between the dominant species in a community and the optimum temperature for foraging (Cerdá et al. [1998]; Angilletta et al. [2007]; Lessard et al. [2009]).

The leaf-cutting ants appear to be a suitable group to test the hypothesis associated with global warming. This group is defined on the basis of its feeding habits and consists of *Atta* and*Acromyrmex* genera, the dominant herbivores of the Neotropics that can harvest up to 17% of the primary productivity (Hölldobler and Wilson [1990]). They have harvesting patterns associated with ranges of the following: specific temperatures for each species and their activity ceases outside those ranges (Hölldobler and Wilson [1990]; Pilati and Quirán [1996]; Bollazzi [2008]; Bollazzi and Roces [2011]); temperatures for the close regulation of cultivating symbiotic fungi - the protein source of the colony (Bollazzi and Roces [2010]; Mueller et al. [2011]); and territorial behavior (Hölldobler and Lumsden [1980]) and aggressive behavior between colonies, associated with the *dear* *enemy* phenomenon (Fisher [1954]; Dimarco et al. [2010]). In the *Attini* tribe,*Acromyrmex lobicornis* Emery has the widest geographical distribution in Argentina, being found from the north down to the 44°S parallel in the province of Chubut (Kusnezov [1978]; Farji-Brener and Ruggiero [1994]). Although its ample geographical distribution suggests that it harvests a wide variety of plant species, there are few studies on its diet in the different habitats where it is found (Farji Brener and Protomastro [1992]; Pilati et al. [1997]; Franzel and Farji-Brener [2000]). *A. lobicornis* behaves as a generalist species: it forages both monocotyledons and dicotyledons; it uses a wide range of species for cultivating the symbiotic fungus; it is an opportunist, taking advantage of food resources when they are available (spatially and seasonally). It also exploits habitats of low complexity and intense aridity, which has given it adaptive plasticity favoring its wide geographical distribution (Pilati et al. [1997]; Claver [2000]). Variations may be observed in the harvesting patterns of the species linked to temperature, depending on the geographic locality, although they are close to those generally cited for ants, at 10°C to 40°C (Hölldobler and Wilson [1990]). Harvesting patterns in relation to soil temperature were recorded in two studies and the temperature ranges observed were 10°C to 40°C (Claver [2000]) and 17°C to 28°C (Nobua Behrmann et al. [2010]), in both cases in the desert in the center of the Monte in the province of Mendoza, Argentina. On the other hand, in a study undertaken in the province of La Pampa, harvesting activity of *A. lobicornis* was seen over an ambient temperature range of 11.7°C to 22.6°C (Pilati and Quirán [1996]). In all cases, peaks of activity were recorded during spring and at the beginning of autumn, whereas activity ceases almost completely during the winter. Moreover, in summer their foraging behavior changed from diurnal to nocturnal harvesting, always within the determined temperature limits and the same as those recorded for daytime (Pilati and Quirán [1996]; Claver [2000]; Nobua Behrmann et al. [2010]). Likewise, in a study on*Acromyrmex heyeri* (Bollazzi [2008]), in anthills and underground, it can be seen how the soil and air temperature vary during the winter, autumn, and summer. There is hardly any difference between the maximum soil and air temperatures during the three seasons, but the minimum temperature of the air is 5°C to 7°C lower than that of soil over the whole period.

This study aims to increase the knowledge of two aspects of the biology of *A. lobicornis* in a controlled environment and of their interactions with the temperature. Each aspect of the effect of variation in temperature was analyzed separately: (a) to evaluate the harvesting activity, by offering different items normally harvested by the species in the area, under two treatments, with a difference in mean temperature of approximately 4.5°C (scenario A2); (b) to measure the magnitude of the modification in the locomotor activity (walking speed) of the workers, in function of a gradual increase in temperature (6°C to 32°C).

METHODS

Study Area

The field work was conducted on land belonging to the Departamento de Agronomía, with coordinates of 38°41′ S; 62°15′ W (Figure 1). Laboratory activities were carried out in the facilities of the Ant Study Group - CERZOS-CONICET, Universidad Nacional Del Sur (UNS), Altos del Palihue 8000, and Bahía Blanca.

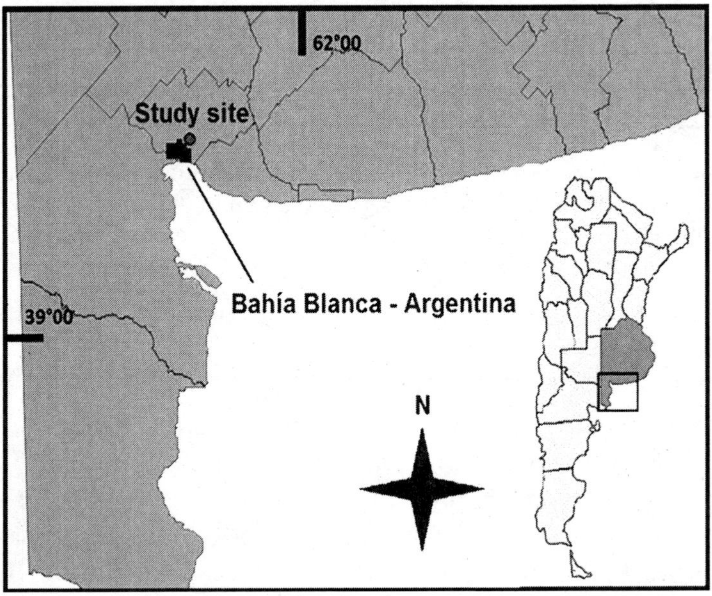

Figure 1: Study area where ant nests were collected. Bahía Blanca, Argentina (38° 41′ S; 62° 15′ W).

Sampling and observations were conducted prior to the study in order to characterize the study area and to complement the theoretical information already collected.

Climate

Historical climatic data from Bahía Blanca, recorded in the meteorological station: 877500 (SAZB), at the city airport (38.73° S; 62.16′ W), at an altitude of 75 m (Aeródromo de Bahía Blanca[2011]). The means for the last 40 years (1970 to 2010): mean annual temperature, 15.4°C; maximum mean annual temperature, 22.8°C; minimum mean annual temperature, 9.0°C; annual precipitation - rain and melted snow, 711.1 mm; mean annual wind speed, 23.3 km/h; and climatic data of 2011 are shown in Table 1.

Table 1: Seasonal means for 2011

Season	T med	T max	T min	Ppm	Ws (med)	Ws (max)	D. Dom
Autumn	14.18	28.20	1.58	30.90	7.03	55.13	NNW
Winter	10.00	22.50	−3.80	6.70	6.40	55.50	NNW
Spring	14.05	27.80	−2.75	18.80	7.15	58.75	NNW

Values of mean, maximum, and minimum temperature (T med, T max, and T min) were measured in degrees Celsius, precipitation in millimeter, and wind speed (Ws) in kilometer per hour. D. Dom is the direction of the prevailing wind. Tizón *et al. Zoological Studies* 2014 53:40, doi: 10.1186/s40555-014-0040-4

Soil

The study site is in the lower valley of the Naposta stream, which is located at the distal end of the western plain surrounding the Ventania range. The valley was shaped during a humid paleoclimate whereby the slopes are very extensive in comparison to the current stream bed. The slope of the sector is 4% to 5%, simple, and running 150 m in a SW-NE direction. A description of a soil profile at 38° 41′ S; 62° 15′ W shows that materials are derived from wind-eroded sediments of moderately coarse texture. The morphology of the profile shows a succession of horizons (A)-C-Ck1-2Ckm (layer of coarse calcareous soil 'tosca' or petrocalcic horizon) and it is classified as a moderately deep coarse loamy, thermic, petrocalcic Paleustol. Its main limitations are related to its low chemical fertility (1% total OM), coarse texture (sandy-loam), poor structure, low water holding capacity (70 mm in a depth of 1 m), a limited effective depth (tosca is found between 0.30 and 1 m), and high susceptibility to water and wind erosion (Bravo, personal communication).

Experimental Design

Foraging Activity

The leaf-cutting ant *A. lobicornis* behaves as a generalist species, as it uses both monocotyledons and dicotyledons: as polyphagous, for the wide range of species used to grow the fungus, and as an opportunist, for its use of the available food sources and for adapting its patterns of activity to the seasonal conditions (Pilati and Quirán [1996]; Pilati et al. [1997]; Claver [2000]; Franzel and Farji-Brener [2000]).

Construction of Artificial Nests

Glass tanks, 30-cm wide and 40-cm deep, were used for the nests. The top of the tank was closed with wood fiber board of the same dimensions as the tank, in which a rectangular door (15 cm × 25 cm) was cut. The door was closed using a stainless steel mesh (100 μm) sealed with a strip of 1 cm Velcro (stuck with contact cement to the perimeter of the mesh and on the perimeter of the opening in the chapadur) so that the mesh could be removed. Two plastic containers with lids were placed in each nest, one to serve as a feeder (14.3 cm × 14.3 cm) and the other for garbage (12.4 cm × 8.6 cm). In both cases, a tube (15-cm long and 3/4 in diameter, graduated every 1 cm) was placed as an entrance to the containers in order to measure the speed of the ants when foraging. Infrared 100-W lamps, regulated with a timer, were used to reach the required temperature, in successive sequences of 2 h on/2 h off.

Nest Collection

In total, 22 nests were found at the study site of which 6 were empty. About 400 individuals were collected at random from 16 nests without making any caste distinction, so the percentage of workers collected was at random. No queens, larvae, eggs, or symbiotic fungi were found (with the exception of three nests where fungus was found) and therefore these elements were not included in the samples. A sample of earth and nesting material (approximately 2 kg) was taken from each nest so that the workers could build new (artificial) nests. About 2 g of symbiotic fungus from a larger nest near to the study area was added to each nest. The sample of fungus was previously wetted with distilled water to decrease the pheromones associated with the original colony to some extent, thereby avoiding any possible rejection of the fungus (Lambardi et al. [2007]; Richard et al. [2007]; Poulsen and Boomsma [2005]). Twenty days later, we replenished the populations of the nests with between 100 and 400 workers, depending on the mortality rate in each nest. At the same time, we added a second sample of symbiotic fungus, 2 g for each nest, to make sure that there was some fungus present. The observations were started a week later.

Forages

The selected items for the experiment were the same plant species used by *A. lobicornis* where it normally feeds in its natural habitat (Table 2).

Table 2: Details of ten forage items as presented to the ants and their state of maturity

Treatments

Species	**Piece**	**State**
Cynodon dactyl on	Leaf	Mature
Diplotaxis tenuifolia	Leaf	Mature
Olea europaea	Leaf	Mature
Olea europaea	Leaf	Shoot
Piptochaetium napostaense	Leaf	Mature
Prunus cerasifera	Petal	Mature
Rhynchosia diversifolia	Stem with leaves	Mature
Schinus molle	Leaf	Mature

Schinus molle	Leaf	Dry
Solanum sisymbriifolium	Leaf	Dry

Tizón *et al.*

Tizón *et al.* *Zoological Studies* 2014 53:40, doi: 10.1186/s40555-014-0040-4

Treatments

Two treatments were used to study the interactions between a given range of temperatures and the foraging activity of *A. lobicornis*:

Tamb: ambient temperature

Tamb+: ambient temperature +4.5°C

Both treatments were subjected to natural variations in temperature (daily and seasonal), which define the temperature ranges in each treatment (range = mean temperature of the treatment [±] *n* degrees). The experiment was performed during spring 2011 in an observation chamber, which is an 'open' room with temperature fluctuations similar to the exterior temperature. The difference of 4.5°C between treatments is measured as the amplitude between the mean temperatures in each treatment. The range of temperatures for each treatment was recorded once a minute throughout the experiment. A Campbell 21X data logger (Campbell Scientific Inc., North Logan, UT, USA) connected to a PC was used to record the temperatures.

Each treatment was applied to eight experimental units (e.u.) selected at random. Ten forage items were offered and a photoperiod of 12/12 h was used (Valderrama et al. [2006]).

The rate of foraging under each treatment was measured and also for each forage item individually, by counting the units of each item harvested from the feeder. Twenty portions of each item were provided. Every 3 days, the forage items were replaced with fresh material and the number of units of each forage type harvested was counted. In total, ten measurements were taken (repetitions). After each repetition, the nests were wetted using a spray bottle, with approximately 75 ml water per nest, which gave a relative humidity of between 60% and 65% (90% after wetting) for nests in the Tamb treatment and 40% to 50% (85% after wetting) for Tamb + .

The graduated tubes at the entrance of each feeder, previously described, were used to measure the walking speed in the harvesting trial. Measurements were made of the walking speed of unladen workers in centimeter per second without making any distinction of subcaste. The walking speed of five workers per nest was measured (these values were then averaged to give a walking speed per nest for each repetition ($n = 10$)), and the procedure was repeated at the same time as the replenishment of material and measurement of harvest per forage item.

Temperature of the Treatments

The range of temperatures in the observation room for the Tamb treatment was between 19.5°C and 22.9°C, with a mean temperature for the whole period of 21.7°C; for the Tamb + treatment, it was between 22.2°C and 30.1°C, with a mean temperature for the whole trial of 26.4°C. The variation in the mean temperature for each treatment during the trial and also the maximum and minimum temperatures reached in each treatment were recorded.

Both treatments were affected by seasonal variation in temperature, however, the mean temperature of the heat treatment was always higher than the mean for the treatment at ambient temperature, with an average temperature range of 4.5°C (±0.75°C).

Loco motor Activity

In order to establish the effect of temperature on walking speed, the speed was measured at ambient temperatures between 6°C and 32°C, which are below and above the activity range of 11.7°C to 28°C cited for *A. lobicornis* (11.7°C to 22.6°C Pilati et al. [1997] and 17°C to 28°C Nobua Behrman et al. [2010]). A brood chamber was used to reach the desired temperature in each case.

Individuals were collected from a colony at the study site. The nest was built from a 20-l plastic container, closed on top by a perforated lid to allow ventilation, and two 1-l containers were added (as feeder and garbage), connected to the main container by graduated (every 1 cm) tubes 15-cm long, with 3/4 in diameter. The new colony included a queen and symbiotic fungus from the original colony. A 12/12 h photoperiod was used (Valderrama et al. [2006]), and the colony was fed on fresh petals of *Pelargonium hortorum* (collected the same day) during the trial.

The walking speed of unladen workers (cm/s) was measured at a range of temperatures between 6°C and 32°C, at 2°C intervals, taking 30 measurements per temperature. This was carried out by day with an aclimatization period of 24 h between one measurement and the next.

Statistical Analysis

Data for the quantity and speed of foraging and forage preference were analyzed with non-parametric models, after carrying out normality homoscedasticity tests - Shapiro-Wilks and de Levene, respectively - required for parametric models (Balzarini et al. [2008]). Tests of Wilcoxon with two tails were performed to compare the results obtained of quantity (20 portions) and harvesting speed (cm/s) in function of the different temperatures (21.7°C and 26.4°C). In order to compare forage preference (ten items), a non-parametric analysis of variance to one-way classification proposed by Kruskal-Wallis and comparisons of pairs between the means of the ranges of treatments (=0.05) were performed. This test allows the comparison of more than two distributions or more expected distributions without the necessity of complying with the assumption of normal distribution for the terms of error (Balzarini et al. [2008]).

Analysis of the ants' walking speed (cm/s), in function of a gradual increase in the ambient temperature (6°C to 32°C), was performed using simple polynomial regressions (grades 2, 3, and 4). A logarithmic transformation to base 10 was previously applied to the data to comply with the supposed requirements of the model. The results are presented without transformation to facilitate their interpretation.

RESULTS

Foraging Activity

The rate of harvesting was 33% higher under the treatment of ambient temperature +4.5°C (Tamb+) ($W = 95$, $p = 0.03$) when all foraged items were considered together. The patterns of harvesting by *A. lobicornis* showed variable levels of preference between the different items, both in the ambient temperature treatment (Tamb) ($H = 54.7$, $p = 0.0001$) and in the Tamb + treatment ($H = 52.5$, $p = 0.0001$). In both treatments, *Piptochaetium napostaense* (leaf) was the least harvested item, and *Prunus cerasifera* (petal) the most harvested (Figures 2 and 3).

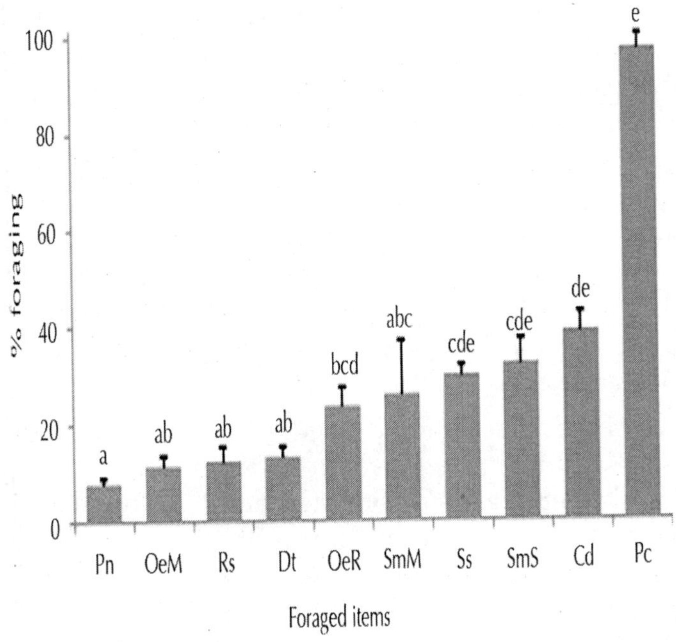

Figure 2: Treatment at ambient temperature. Foraged items: Cd, *Cynodon dactylon*; Dt, *Diplotaxis tenuifolia*; OeM, *Olea europaea*(mature); OeR, *Olea europaea* (shoot); Pn, *Piptochaetium napostaense*; Pc, *Prunus cerasifera*; Rs, *Rhynchosia senna*; SmM, *Schinus molle* (mature); SmS, *Schinus molle* (dry); Ss, *Solanum sisymbriifolium*. Bars with different letters show significant differences ($p < 0.05$).

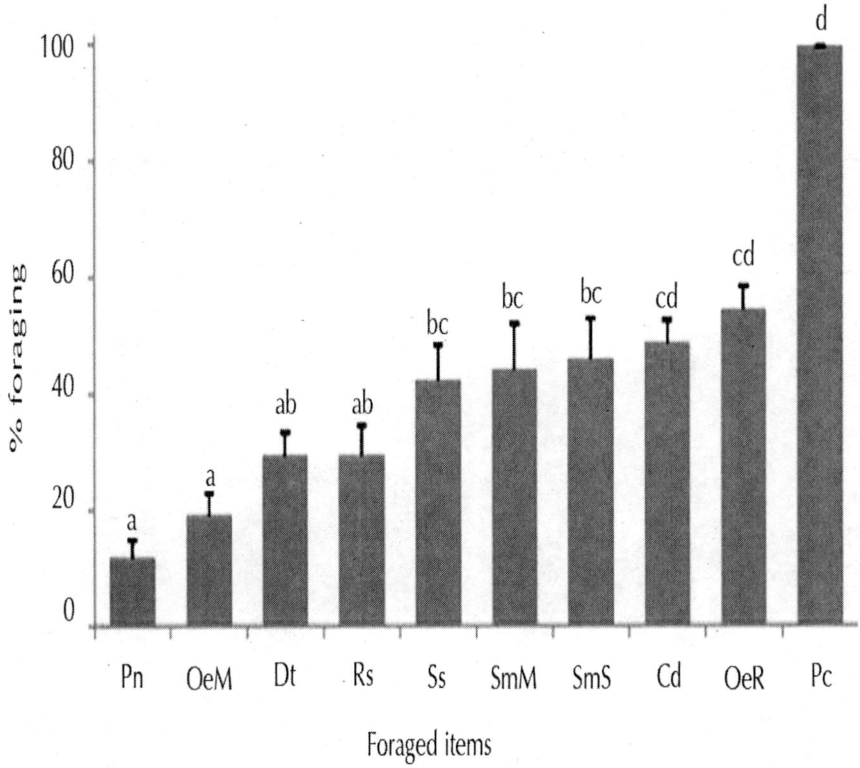

Figure 3: Treatment at ambient temperature +4.5°C. Foraged items: Cd, *Cynodon dactylon*; Dt, *Diplotaxis tenuifolia*; OeM, *Olea europaea* (mature); OeR, *Olea europaea* (shoot); Pn, *Piptochaetium napostaense*; Pc, *Prunus cerasifera*; Rs, *Rhynchosia senna*; SmM, *Schinus molle* (mature); SmS, *Schinus molle* (dry); Ss, *Solanum sisymbriifolium*. Bars with different letters show significant differences ($p<0.05$).

All ten forage items used in the study were harvested in greater proportion under heat treatment (Figure 4A,B,C,D,E,F,G,H,I,J), and significant differences between treatments were found in four of them: *Diplotaxis tenuifolia* ($W=95$, $p=0.0026$) (Figure 4B); *Olea europaea* shoot ($W=99$, $p=0.0003$) (Figure 4D); *Rhynchosia senna* ($W=90$, $p=0.0207$) (Figure 4G); *Schinus molle* mature ($W=87$, $p=0.0457$) (Figure 4H). On the other hand, there were no differences between treatments for two of the forage items: *Piptochaetium napostaense* ($W=69$, $p=0.9394$) (Figure 4E); *P. cerasifera* ($W=68.5$, $p>0.9999$) (Figure 4F).

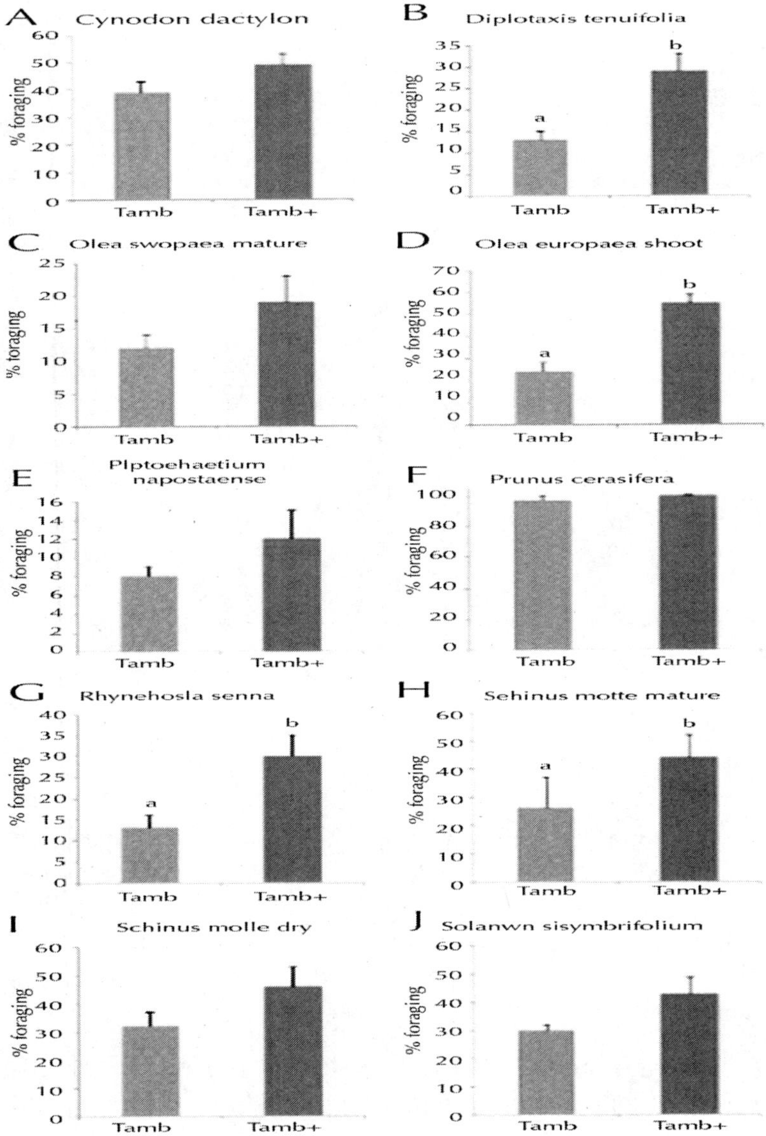

Figure 4: Harvesting percentages of the ten forage items and differences in foraging rate. Harvesting percentages of the ten forage items and differences in foraging rate (in proportions) under each treatment for each item, with the associated SE in each case.(A)*Cynodon dactylon*.(B)*Diplotaxis tenuifolia*. (C)*Olea europaea* mature.(D)*Olea europaea* shoot.(E)*Piptochaetium napostaense*.(F)*Prunus cerasifera*.(G)*Rhynchosia senna*.(H)*Schinus molle* mature. (I) *Schinus molle* dry. (J)*Solanum sisymbriifolium*. The bars with different letters

indicate significant differences ($p < 0.05$). Harvesting percentages of the ten forage items, and differences in foraging rate (in proportions) under each treatment for each item, with the associated SE in each case. The bars with different letters indicate significant differences ($p < 0.05$).

Two items were offered in two different states: *O. europaea* (mature and shoot) and *S. molle* (mature and dry). The shoots of *O. europaea* were harvested in greater proportion than the dry material in both treatments (Figures 2 and 3), although the difference was only statistically significant ($H = 52.5$, $p = 0.0001$) under the Tamb + treatment (Figure 3). However, the dry material of *S. molle* was harvested in greater proportion than the mature in both treatments (Figures 2 and3), although it was only statistically significant ($H = 54.7$, $p = 0.0001$) under the Tamb treatment (Figure 2).

Loco motor Activity

In the foraging speed trial, higher walking speed was observed in the Tamb + treatment (2.40 cm/s) than in Tamb (1.89 cm/s) ($W = 99$, $p = 0.0003$) (Figure 5).

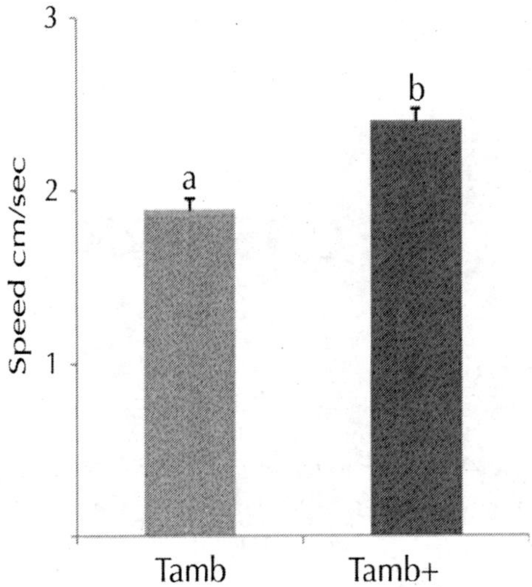

Figure 5: Walking speed of the workers under both treatments Tamb and Tamb + 4.5°C. The different letters indicate the significant differences ($p < 0.05$).

The mean speed of the workers gradually increased in function of an increase in the temperature, up to a maximum speed of 2.85 cm/s at 22°C, and then it gradually decreased up to 32°C (Figure 6).

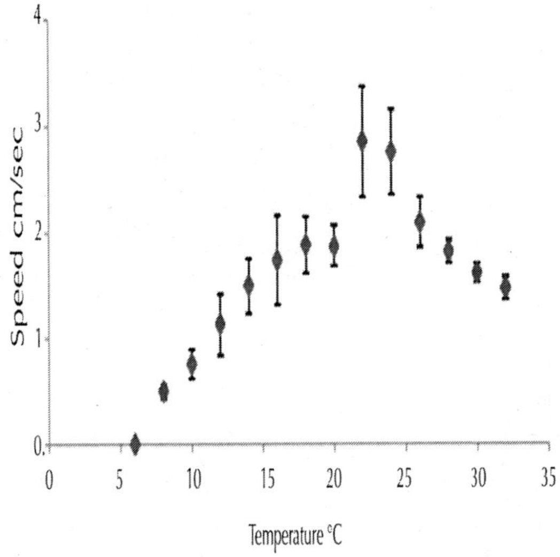

Figure 6: Walking speed of the workers with a temperature between 6°C and 32°C. The lines indicated SE in each bar ($R^2 = 0.9083$, $y = -0.0003x^3 + 0.0077x^2 + 0.1155x - 0.8684$).

DISCUSSION

Foraging Activity

The leaf-cutting ant *A. lobicornis* is considered as a generalist species in regard to its choice of forage plants (Pilati et al. [1997]; Claver [2000]; Franzel and Farji-Brener [2000]). However, if the percentages for each item harvested in our study over the total biomass foraged in both treatments are taken into account, we can define a variable level of preference for each forage item harvested.

The results observed, which show a selective rate of harvesting of certain items, can be explained by two hypotheses. The first is related to chemical defenses present in certain plants that have a variable content of components which are toxic or disagreeable to ants (tanins, alcaloids) (Farji Brener and Protomastro [1992]; Franzel and Farji-Brener [2000]; Caffarini et al. [2008]). This assumes that the presence of components that are disagreeable to ants in a forage item would lead to a lower harvesting rate of this item. The second hypothesis refers to the balance that exists in every plant species between components that are degradable and non-degradable by the symbiotic fungus (Pilati et al. [1997]; Gomes de Siqueira et al. [1998]; Franzel and Farji-Brener [2000]; Abril and Bucher [2004]; Bucher and Marchesini [2004]). The nutritional interactions between the fungus (*Leucoagaricus*) and the ants are largely unknown, although it appears that there may be enzymatic contributions from both symbiotic partners in the degradation and processing of the collected plant material (Richard et al. [2005]; Erthal et al. [2004]). Plant components which symbiotic fungi are not capable of degrading are the following: hemicellulose, cellulose, lignin, and silicon; and the degradable components are the following: soluble carbohydrates (cotton and xylose), phenols, and pectins (according to the enzymatic activity of the fungus) (Gomes de Siqueira et al. [1998]; Abril and Bucher [2004]; Bucher and Marchesini [2004]; De Fine Licht et al. [2009]). This hypothesis assumes that the forage items with less compounds that are not degradable by the fungus would have higher harvest rates. The first hypothesis does not explain the case of the *S. molle* item, since the level of preference was intermediate due to its high content of essential oils that have a repellent and toxic effect (Iannacone and Lamas [2003]; Werdin González et al. [2008]). The second hypothesis seems to explain better the differential preference between different forage items. This hypothesis, for example, confirms that when two forage items *Cynodon dactylon* and *P. napostaense* are compared, they show high and low levels of preference, respectively. There are different assays for qualifying the forage value of a plant, such as NDF analysis (Neutral Detergent Fiber), which indicates the percentage of insoluble components from plant cell walls present in a solution of neutral detergent (basically cellulose, hemicellulose, lignin, and silicon). The values recorded for these species are up to 76% for *C. dactylon* and up to 82% for *P. napostaense* (Van Soest [1994]; Juárez Reyes et al. [2009]). Moreover, it should be noted that *C. dactylon* is described as a tropical grass and *P. napostaense*as temperate and, in general, tropical grasses have higher NDF values than those of temperate climates since more metabolic energy is invested in the supporting and conducting tissues than in reserve tissue. These values support the idea that the higher the content of components non-degradable by symbiotic fungi, the lower the level of harvesting preference by the workers. On the other hand, the item with the highest harvesting rate was petals of *P. cerasifera* that has a low content of cellulose, lignin, and hemicellulose which are abundant cell wall components, e.g., in supporting and conducting tissues, and are more developed in stems and leaves (Bidwell [1993]). Leaf-cutting ants show foraging preference for young plant tissues, e.g., the apical meristems of young plants, either due to the low content of cellulose and lignin or defense chemicals (Saluso and Xavier [2010]), which we could also confirm in our study, as seen in the higher harvesting rate of *O. europaea* in the form of leaves from a shoot (11%) in comparison to mature leaves of the same species (4.3%). This phenomenon is in agreement with that observed in the quality analysis of forage plants. The content of proteins and insoluble fibers (cellulose, hemicellulose, and lignin) varies in each plant according to its state of maturity (phenological stage). The young plants possess a high content of proteins and low content of insoluble fibers in comparison to mature plants in which the protein content is lower and the content of insoluble fibers associated with conducting and supporting tissues is increased (Van Soest [1994]; Juárez Reyes et al. [2009]). This high content of proteins in leaves and young organs, compounds that are degradable by the symbiotic fungus, and the low insoluble fiber content that the fungus is not capable of degrading successfully might explain why the workers harvest these types of forage items preferentially.

It should be clarified that all the forage plants had been harvested by ants in the field during the study. Moreover, the length of the trial was too short to assume any positive signal on the part of the fungus as feedback from the workers on what they harvested. In the leaf-cutting ants in particular, the environmental element that activates a 'positive signal' for harvesting originates in the symbiotic fungus cultivated by the colony and there is a positive correlation between the plant materials harvested by the workers and the capacity of the fungus to metabolize it (Saverschek and Roces [2011]).

When we analyze the harvesting rate by treatment, we can see that the magnitude of harvest for all forage items was greater under the +4.5°C (Tamb+) ambient temperature treatment. These observations are in agreement with the harvesting patterns of *A. lobicornis* and its relationship with temperature observed by Pilati and Quirán ([1996]). In the Tamb + treatment, with an ambient temperature range of 22.2°C to 30.1°C, it was seen that the harvesting rate of all items together was 1/3 greater than for the ambient temperature range of 19.5°C to 22.9°C. Harvesting patterns have been seen for variable temperature ranges for this species: at an ambient temperature of 11.7°C to 22.6°C (Pilati and Quirán [1996]) and a soil temperature of 10°C to 40°C (Claver[2000]) and 17°C to 28°C (Nobua Behrmann et al. [2010]). A temperature range appears to be a variable that predicts the foraging activity in the best way when the harvesting activity of a species is defined *a priori. A. lobicornis* is an opportunist leaf-cutter that harvests a large variety of plant species (Pilati and Quirán [1996]; Pilati et al. [1997]; Claver [2000]; Franzel and Farji-Brener[2000]) and it does so under specific climatic conditions (Pilati and Quirán [1996]; Quirán and Pilati [1998]; Claver [2000]). Studies that analyze competitive interactions between species of ants that forage at the same site point out that they compete for a temperature range that is optimum for their foraging activity (Cerdá et al. [1998]; Lessard et al. [2009]). If we define a harvesting temperature range for *A. lobicornis* of 11.7°C to 40°C, as cited in different places, we can see that the Tamb + treatment lies between the mean temperature of this range (25.9°C) and the upper limit (40°C). The results obtained in our study assert that the maximum harvesting rate occurs at the highest temperatures within the optimum temperature range (Pilati and Quirán [1996]). The highest foraging rate in the higher temperature treatment can be attributed to an effect of the harvesting activity within the upper portion of the optimum harvesting temperature range cited for the species.

Comparing both treatments (Tamb and Tamb+), in the analysis of the rate of harvesting for each forage item, three situations were observed: (1) *P. napostaense* and *P. cerasifera* were the least and most harvested items, respectively, and there was no variation in the foraging rate due to the difference in temperature; (2) The difference between treatments was not significant for four of the items of intermediate preference; (3) For the remaining four items of intermediate preference, there was a significant difference between treatments. It can be said *a priori* that the items of intermediate preference are harvested differentially according to the ambient temperature; but for the highest and lowest levels of preference, no link was detected between the harvesting rate and the variation in temperature. However, if we take the percentages of each item harvested into account for both treatments together and we put the percentages of foraging over the total biomass harvested into order, we can see the exceptions: *D. tenuifolia* and *O. europaea* (shoot) that correspond to 5.8% and 11% of the total biomass harvested, respectively, showed differences under both treatments. If we define a gradient of foraging preference for the ten forage items, we find that *D. tenuifolia* and *O. europaea* (shoot) are placed in the lowest and highest preference levels, respectively. On the other hand, if we consider *S. molle* (mature) and *Solanum sisymbriifolium* that correspond to 9.7% and 10.1% of the total biomass harvested, respectively (a more intermediate level of preference), the differences in harvesting rate between both treatments were not significant. Moreover, if we consider the difference in harvesting rates between treatments for each item (in percentages), the value that separates the three groups or situations previously described is only 2 or 3 units. In conclusion, it would be correct to assert that the differences in the rates of harvesting for each forage item under each treatment, whether significant or not, may be an artifact of the experiment and a chance effect, but it can also be stated that the forage rate is greater when there is an increase in temperature when referring to all the items together, although this difference is not seen for each individual foraged item.

Lastly, when we look in particular at the two plant species that were presented in two different forms, *O. europaea* (shoot and mature) and *S. molle* (mature and dry), in both cases, the harvesting differential at different temperatures may be in response to the needs of the colony. *O. europaea* (shoot) was harvested in greater proportion in both treatments, being only statistically significant under the higher temperature treatment. This might be due to the behavior of the workers, as the colony expands when the temperature is elevated and nutritional needs are prioritized (Pilati et al. [1997]). On the other hand, in the case of *S. molle*, in which the dry state was harvested in greater proportion at lower temperatures, it may be that the ants were harvesting chaff for the construction of an anthill to protect the nest during the low winter temperatures, which occurs in autumn when the ambient temperature starts to drop (Pilati et al. [1997]; Bollazzi [2008]; Perez [2009]).

Loco motor Activity

Two trials on the walking speed of the workers were performed in this study. The first was carried out during the trial on foraging activity when the walking speed was measured under two temperatures separated by an average thermic amplitude of 4.5°C. The second was designed with the measurement of walking speed as the only objective, when the way in which the walking speed changes under a gradient of ambient temperature (6°C to 32°C) was evaluated. In the first trial, it was shown that a thermic amplitude of 4.5°C is sufficient to increase the walking speed of the workers by more than 20%. The walking speed of the workers of *A. lobicornis* increases as the ambient temperature increases, up to a given maximum, and then it starts to drop as the ambient temperature continues to rise. This phenomenon has also been seen in other ant species in a similar pattern, although the maximum speed may be reached at different temperatures (Angilletta et al. [2008]; Bollazzi and Roces [2011]). Ants from different species or colonies that forage at the same site compete for resources, as well as for the ambient temperature in which they carry out their foraging activity, the dominant species foraging within the most suitable ranges of temperature for the activity (Cerdá et al. [1998]; Lessard et al. [2009]). In this way, an equilibrium is established when different groups forage at different times of the day, since each group is active within a determined temperature range and small variations of a few degrees do not alter the foraging activity (Chambers [2011]). *A. lobicornis* forages in a temperature range of 10°C to 40°C either by day or by night (Pilati and Quirán [1996]; Claver [2000]; Nobua Behrmann et al. [2010]). The results obtained show that for this species that harvests within a range with a 30°C amplitude, a difference of 4.5°C is a sufficient thermic amplitude to make a significant change in the walking speed of the workers. In the second walking speed trial, with similar sizes of workers, the walking speed was seen to increase up to a maximum between 22°C and 24°C, after which it began to drop down. As has already been said, this pattern of walking speed, which can be associated with foraging activity, is similar for different species, but the maximums are reached at different temperatures. This behavior might be explained by two hypotheses. The first refers to the thermic tolerance range of each species. However, the second hypothesis assumes that the competitive interactions between different groups of ants at the same site define the temperature range within which each species forages and so the walking speed is associated with this activity (Cerdá et al. [1998]; Lessard et al. [2009]). On the basis of the first hypothesis, we can mention the example of thermic tolerance and the range of activity of two ant species, *Atta sexdens* and *A. lobicornis*, which have similar thermic tolerance ranges. However, the former species reaches its maximum walking speed between 6°C and 8°C higher than the latter species (Pilati and Quirán[1996]; Claver [2000]; Angilletta et al. [2007], [2008]; Nobua Behrmann et al. [2010]). The thermic tolerance differential per species does not explain satisfactorily why each group of ants reaches its maximum walking speed at different temperatures. So the second hypothesis appears to give a possible explanation for the differences between the species, but it cannot be put to test without observing the patterns of foraging of the ant species present in the study area and how they relate in respect to the temperature and to each other.

The efficiency with which the workers carry out the foraging activity might be related to the size of item that each worker carries (Quirán and Steibel [2001]; Bollazzi and Roces [2011]). Although we could also relate it to the walking speed of the workers, as the foraging efficiency of a species not only depends on the size of item carried by the workers but also on the speed at which it is transported to the nest. In this study, we were able to see that as the ambient temperature increased the workers' harvesting for more forage (1/3 more plant biomass for a difference of 4.5°C), and moreover they moved at a greater speed up to a maximum at a determined temperature (22°C to 24°C) and then began to move more slowly from then onwards. Therefore, it could be said that an increase in the ambient temperature up to around 25°C increases the harvesting efficiency of *A. lobicornis*. Since the temperature in summer is often above this level but it is not reached during the winter, global warming will have the tendency to make a significant impact on the harvesting activity of this species in late autumn and early spring.

The speeds differed according to the experiment (foraging and walking activities); when the temperature was around 22°C, the speed was higher in the walking experiment and at 26°C it was the opposite. This can be explained by the fluctuation in ambient temperature in the foraging experiment, similar to the outside (maximum day temperature and minimum night temperature). This amplitude allows the workers to rest when the temperature is high. On the other hand, in the speed experiment, the workers were in a chamber at a constant temperature each day (day and night).

Ants are ectothermic and they are directly affected by the ambient temperature when outside their nests (Bale et al. [2002]; Angilletta et al. [2002]; Deutsch et al. [2008]). However, they can actively control the microclimate inside their nests by alleviating, isolating, or holding back the heat with their characteristic structures (Quirán and Pilati [1998]; Kleineidam et al. [2001]; Bollazzi and Roces [2007]). So, species of *Acromyrmex* have developed specific mechanisms to control the microclimate of the nest, creating thermic stability which allows them to live under extreme conditions of humidity and temperature (Quirán and Pilati [1998]). The workers can move the fungus gardens to the most appropriate areas in the nest and change the architecture which has an effect on the ventilation (O_2-CO_2 exchange) (Kleineidam et al. [2001]; Bollazzi and Roces[2007]) and stir up the soil around the nest which increases its porosity. Moreover, as the soil density is decreased, it favors the infiltration of water and gaseous exchange (Quirán and Pilati [1998]; Kleineidam et al. [2001]). In both experiments undertaken in this study, the capacity of the ants to move the fungus and to remodel the nest was limited; however, the aim of this study was principally to measure the effects of temperature on activities outside the nests.

CONCLUSIONS

The projections generated for the climatic model by CIMA/CONICET show an increase in temperature of between 2°C and 4.5°C in the next 80 years for the region that extends from the north of Argentina to the north of Patagonia. The geographic distribution of *A. lobicornis* within Argentina is from the north of the country to the 44°S parallel. In our study we observed that an increase of 4.5°C in the ambient temperature was sufficient to make an impact on the activity of leaf-cutting ants:

Higher rate of harvesting (33% higher)

Significantly faster walking speed (27% higher)

If we consider all the facts discussed in this study together (plasticity in the diet of leaf-cutting ants, harvesting a wide variety of plant species - monocotyledons and dicotyledons; a preference for harvesting young leaves with a low percentage of cellulose, hemicellulose, lignin, and waxes; highly capable of becoming adapted to different sites and climates; behavioral habits that allow it to become adapted to a variation in ambient temperature; and a citation of the species in the N of Patagonia where the ambient temperature is a limiting factor for the species), in the face of a global warming scenario, with an increase in temperature of between 2°C and 4.5°C for this area, it can be seen that this species might become a pest of greater magnitude (Perez [2009], although currently described as a moderate pest by Fowler et al. [1986]), foraging not only on crops and forest plantations but also with increased foraging pressure on native plant species. It could also become more competitively aggressive to other animal species that forage in the same area as *A. lobicornis*.

AUTHORS' CONTRIBUTIONS

RT wrote the preliminary draft of this manuscript, carried out the design and coordination of the study. RT and JPW carried out the experiment. DVP gave advice in the statistical analysis and revised the manuscript. All authors have read and approved the final version of the manuscript.

ACKNOWLEDGEMENTS

We are thankful to Omar Elia and Osvaldo Fernández for the materials (glass tanks) for the experiments. We are grateful to Aaron Ellison PhD for first project ideas. We thank the reviewers for substantially improving the manuscript. This study was supported by the CERZOS CCT-Bahía Blanca CONICET.

REFERENCES

1. Abril AB, Bucher EH (2004) Nutritional sources of the fungus cultured by leaf-cutting ants. Appl Soil Ecol 26:243-247
2. (2011) Clima en Bahia Blanca Aerodrome [on line].
3. Angilletta MJ, Niewiarowski PH, Navas CA (2002) the evolution of thermal physiology in ectotherms. J Therm Biol 27:249-268

4. Angilletta MJ, Wilson RS, Niehaus AC, Sears MW, Navas CA, Ribeiro PL (2007) urban physiology: city ants possess high heat tolerance. PLoS ONE 2:e258
5. Angilletta MJ, Roth TC, Wilson RS, Niehaus AC, Ribeiro PL (2008) the fast and the fractalous: speed and tortuosity trade off in running ants. Funct Ecol 22:78-83
6. Bale JS, Masters GJ, Hodkinson ID, Awmack C, Bezemer TM, Brown VK, Butterfield JEL, Buse A, Coulson JC, Farrar J, Good JEG, Harrington R, Hartley S, Jones TH, Lindroth RL, Press MC, Symrnioudis I, Watt AD, Whittaker JB (2002) Herbivory in global climate change research: direct effects of rising temperatures on insect herbivores. Glob Change Biol 8:1-16
7. Balzarini M, Gonzalez L, Tablada M, Casanoves F, Di Rienzo JA, Robledo CW (2008) Infostat. Manual Del Usuario, Editorial Brujas, Córdoba, Argentina.
8. Bestelmeyer BT (2000) the trade-off between thermal tolerance and behavioural dominance in a subtropical South American ant community. J Anim Ecol 69:998-1009
9. Bidwell RGS (1993) Fisiología Vegetal. 1era edición en español. AGT EDTTOR, S.A.México.
10. Bollazzi M (2008) Building behaviour and the control of nest climate in Acromyrmex leaf-cutting ants. Doctoral thesis - Deutscher Akademischer Austauschdienst, Germany.
11. Bollazzi M, Roces F (2007) to build or not to build: circulating dry air organizes collective building for climate control in the leaf-cutting ant *Acromyrmex ambiguus*. Anim Behav 74(5):1349-1355
12. Bollazzi M, Roces F (2010) Control of nest water losses through building behavior in leaf-cutting ants (*Acromyrmex heyeri*). Insect Soc 57:267-273 Publisher Full Text
13. Bollazzi M, Roces F (2011) Information needs at the beginning of foraging: grass-cutting ants trade off load size for a faster return to the nest. PLoS ONE 6(3):e17667
14. Bollazzi M, Kronenbitter J, Roces F (2008) Soil temperature, digging behaviour, and the adaptive value of nest depth in South American species of Acromyrmex leaf-cutting ants. Oecologia 158:165-175

15. Bucher EH, Marchesini V (2004) Herbivory by leaf-cutting ants: nutrient balance between harvested and refuse material. Biotropica 36(3):327-332

16. Buckley LB, Jetz W (2008) Linking global turnover of species and environments. PNAS 105:17836-17841

17. Bucy AM, Breed MD (2006) Thermoregulatory trade-offs result from vegetation removal by a harvester ant. Ecol Entomol 31(5):423-429

18. Caffarini P, Carrizo P, Pelicano A, Roggero P, Pacheco J (2008) Efectos de extractos acetónicos y acuosos de *Ricinus communis* (ricino), *Melia azedarach* (paraíso) y *Trichillia glauca* (trichillia), sobre la hormiga negra común (*Acromyrmex lundi*) effects of acetonic and water extracts of *Ricinus communis*, *Melia azedarach* y *Trichillia glauca* on black common cutting ant (*Acromyrmex lundi*). Idesia 26(1):59-64

19. Cerdá X, Retana J, Cros S (1998) Critical thermal limits in Mediterranean ant species: trade-off between mortality risk and foraging performance. Funct Ecol 12(1):45-55Publisher Full Text

20. Chambers CM (2011) the effects of resource availability and temperature on ants. University of Tennessee Honors Thesis Projects, Knoxville.

21. Claver S (2000) Ecología de Acromyrmex lobicornis (E.) (Hymenoptera: Formicidae) en la Reserva de Biosfera de Ñacuñán, provincia Biogeográfica Del Monte. Tesis de Doctorado - Facultad de Ciencias Naturales y Museo. Universidad Nacional de La Plata, Buenos Aires.

22. Cole BJ, Smith AA, Huber ZJ, Wiernasz DC (2010) the structure of foraging activity in colonies of the harvester ant, *Pogonomyrmex occidentalis*. Behav Ecol 21(2):337-342

23. Cooper BS, Williams BH, Angilletta MJ (2008) Unifying indices of heat tolerance in ectotherms. J Therm Biol 33:320-323

24. De Fine Licht HH, Schiøtt M, Mueller UG, Boomsma JJ (2009) Evolutionary transitions in enzyme activity of ant fungus gardens. Evolution 64(7):2055-2069

25. Deutsch C, Tewksbury JJ, Huey RB, Sheldon K, Ghalambor C, Haak D, Martin PR (2008) Impacts of climate warming on terrestrial ectotherms across latitude. PNAS 105(16):6668-6672

26. Dimarco RD, Farji-Brener AG, Premoli AC (2010) Dear enemy phenomenon in the leaf-cutting ant *Acromyrmex lobicornis*: behavioral and genetic evidence. Behav Ecol 21:304-310
27. Erthal M Jr, Silva CP, Samuels RI (2004) Digestive enzymes of leaf-cutting ants,*Acromyrmex subterraneus* (Hymenoptera: Formicidae: Attini): distribution in the gut of adult workers and partial characterization. J Insect Physiol 50:881-891
28. Farji Brener AG, Protomastro JJ (1992) Patrones forrajeros de dos especies simpátricas de hormigas cortadoras de hojas (Attini, *Acromyrmex*) en UN bosque subtropical seco. Ecotrópicos 5:32-43
29. Farji-Brener AG, Ruggiero A (1994) Leaf-cutting ants (*Atta* and *Acromyrmex*) inhabiting Argentina: patterns in species richness and geographical ranges sizes. J Biogeogr 21:535-543
30. Fisher JB (1954) Evolution and Bird Sociality. In: Huxley J, Hardy AC, Ford EB (eds) Evolution as a Process, Allen, Unwin, London. pp 71-83
31. Fowler GH, Pereira-da-Silva V, Forti LC, Saes NB (1986) Population dynamics of leaf-cutting ants. A brief review. In: Lofgren CS, Vander Meer RK (eds) Fire ants and leaf cutting ants: biology and management, Westview, Boulder. pp 123-145
32. Franzel C, Farji-Brener AG (2000) ¿Oportunistas o selectivas? Plasticidad en la dieta de la hormiga cortadora de hojas *Acromyrmex lobicornis* en el N–O de la Patagonia. Ecología Austral 10:159-168
33. Gomes de Siqueira C, Bacci M, Pagnocca FC, Correa Bueno O, Aparecida Hebling MJ (1998) Metabolism of plant polysaccharides by *Leucoagaricus gongylophorus*, the symbiotic fungus of the leaf-cutting ant *Atta sexdens* L. Appl Environ Microbiol 64(12):4820-4822
34. Hartley S, Lester PJ (2003) Temperature-dependent development of the argentine ant,*Linepithema humile* (Mayr) (Hymenoptera: Formicidae) a degree-day model with implications for range limits in New Zealand. New Zealand Entomologist 26:91
35. Hochochka PW, Somero GN (2002) Biochemical adaptation. Oxford University Press, Oxford.

36. Hölldobler B, Lumsden CJ (1980) Territorial strategies in ants. Science 210:732-739

37. Hölldobler B, Wilson E (1990) the ants. Harvard University Press, Cambridge.

38. Huey RB, Stevenson RD (1979) Integrating thermal physiology and ecology of ectotherms: discussion of approaches. Am Zool 19:357-366

39. Iannacone J, Lamas G (2003) Efecto toxicológicos de extractos de molle (*Schinus molle*) y lantana (*Lantana camara*) sobre *Chrysoperla externa* (Neuroptera: Chrysopidae), *Trichogramma pintoi* (Hymenoptera: Trichogrammatidae) y *Copidosoma koehleri*(Hymenoptera: Encyrtidae) en el Perú. Agri Téc (Chile) 63:347-360

40. (2007) Cambio climático 2007: Informe de síntesis. Contribución de los Grupos de trabajo I, II y III al Cuarto Informe de evaluación Del Grupo Intergubernamental de Expertos sobre el Cambio Climático [Pachauri, R.K. y Reisinger, A. (directores de la publicación)]. IPCC, Ginebra, Suiza.

41. Houghton JT (ed) (2001) Climate change 2001: the scientific basis. Cambridge University Press, New York.

42. Juárez Reyes AS, Cerrillo Sotoa MA, Gutiérrez Ornelasb E, Romero Treviñoc EM, Colín Negreteb J, Bernal Barragán H (2009) Estimación del valor nutricional de pastos tropicales a partir de análisis convencionales y de la producción de gas in vitro. Téc Pecu Méx 47(1):55-67

43. Jumbam KR, Jackson S, Terblanche JS, McGeoch MA, Chown SL (2008) Acclimation effects on critical and lethal thermal limits of workers of the Argentine ant, *Linepithema humile*. J Insect Physiol 54:1008-1014

44. Kaspari M (2004) Using the metabolic theory of ecology to predict global patterns of abundance. Ecology 85:1800-1802

45. Kaspari M, O'Donnell S, Kercher JR (2000) Energy, density, and constraints to species richness: studies of ant assemblages along a productivity gradient. Am Nat 155:280-293

46. Kaspari M, O'Donnell S, Alonso L (2000) Three energy variables predict ant abundance at a geographic scale. Proc R Soc B 267:485-490

47. Kaspari M, Ward P, Yuan M (2004) Energy gradients and the geographic distribution of local ant diversity. Oecologia 140:407-414
48. Kleineidam C, Ernst R, Roces F (2001) Wind-induced ventilation of the giant nests of the leaf-cutting ant Atta vollenweideri. Naturwissenschaften 88(7):301-305
49. Kusnezov N (1978) Hormigas Argentinas: Clave para su identificación. Edición preparada por R. Golbach. Miscelánea 61, Fundación Miguel Lillo.
50. Lambardi D, Dani FR, Turillazzi S, Boomsma JJ (2007) Chemical mimicry in an incipient leaf-cutting ant social parasite. Behav Ecol Sociobiol 61:843-851
51. Lessard JP, Dunn RR, Sanders NJ (2009) Temperature-mediated coexistence in temperate forest ant communities. Insect Soc 56:149-156
52. May ML (1979) Insect thermoregulation. Ann Rev Entomol 24:313-349
53. Mueller UG, Mikheyev AS, Honga E, Sena R, Warrena DL, Solomon SE, Ishaka HD, Cooper M, Millera JL, Shaffera KA, Juenger TE (2011) Evolution of cold-tolerant fungal symbionts permits winter fungiculture by leafcutter ants at the northern frontier of a tropical ant–fungus symbiosis. PNAS 108:4053-4056
54. Nobua Behrmann BE, Milesi FA, de Casenave Lopez J, Farji-Brener A (2010) Patrones diarios y estacionales de actividad de dos especies simpátricas de hormigas cortadoras de hojas en el desierto Del Monte central. IV Reunión Binacional de Ecología (XXIV Reunión Argentina de Ecología y XVII Reunión de la Sociedad de Ecología de Chile), Buenos Aires, Argentina.
55. Nuñez MN (2009) Cambio climático y el clima esperado para Argentina. Mundo Agro 2007. Tecnología y Sustentabilidad. 17–22. SEMA, Buenos Aires.
56. Nuñez MN, Solman S, Cabré MF, Rolla A (2005) Estimación de escenarios regionales de cambio climático mediante el uso de modelos climáticos regionales. Informe Final CIMA/CONICET B-9 Modelos Climáticos Regionales. GEF BIRF PF 51286 AR,
57. Perez SP (2009) Riesgo potencial de la hormiga cortadora de hojas *Acromyrmex lobicornis*para las plantaciones forestales de

la Patagonia. In: Villacide J, Corley J (eds) Manejo Integrado de Plagas Forestales, Laboratorio de Ecología de Insectos EEA, INTA, Bariloche, Cuadernillo N° 6.

58. Pilati A, Quirán E (1996) Patrones de cosecha de *Acromyrmex lobicornis* (Formicidae: Attini) en UN pastizal Del Parque Nacional Lihué Calel, La Pampa, Argentina. Ecología Austral 7:123-126
59. Pilati A, Quirán EM, Estelrich HD (1997) Actividad forrajera de *Acromyrmex lobicornis*Emery (Hymenoptera: Formicidae) en UN pastizal natural semiárido de la provincia de La Pampa (Argentina). Ecología Austral 7:49-56
60. Pol R, Lopez de Casenave J (2004) Activity patterns of harvester ants *Pogonomyrmex pronotalis* and *Pogonomyrmex rastratus* in the central Monte desert, Argentina. J Insect Behav 17(5):647-661
61. Poulsen M, Boomsma JJ (2005) Mutualistic fungi control crop diversity in fungus-growing ants. Science 307:741-744
62. Quirán EM, Pilati A (1998) Estructura de los hormigueros de *Acromyrmex lobicornis* (Hymenoptera: Formicidae) en UN sitio natural semiárido de La Pampa, Argentina. Rev Soc Entomol Argent 57(1–4):45-48
63. Quirán EM, Steibel JP (2001) *Acromyrmex lobicornis* Emery 1887 (Hymenoptera: Formicidae) y el peso de la carga, en condiciones de laboratorio. Gayana (Concepc) 65(2):113-118
64. Richard FJ, Mora P, Errard C, Rouland C (2005) Digestive capacities of leaf-cutting ants and the contribution of their fungal cultivar to the degradation of plant material. J Comp Physiol B 175:297-303
65. Richard FJ, Poulsen M, Hefetz A, Errard C, Nash DR, Boomsma JJ (2007) The origin of chemical profiles of fungal symbionts and their significance for nestmate recognition in*Acromyrmex* leaf-cut- ting ants. Behav Ecol Sociobiol 61:1637-1649
66. Saluso A, Xavier L (2010) Caracterización del daño producido por la "hormiga negra común" en plántulas de girasol. INTA - EEA Paraná Actualización Técnica N° 2 – Maíz, Girasol y Sorgo, pp 101–103

67. Sanders NJ, Lessard JP, Fitzpatrick MC, Dunn RR (2007) Temperature, but not productivity or geometry, predicts elevational diversity gradients in ants across spatial grains. Global Ecol Biogeog 16:640-649
68. Saverschek N, Roces F (2011) Foraging leafcutter ants: olfactory memory underlies delayed avoidance of plants unsuitable for the symbiotic fungus. Anim Behav 82:453-458
69. Storch D (2010) Thermal and energetic controls on biodiversity. In: Gordon Research Conference on Metabolic Basis of Ecology. University of New England in Biddeford,
70. Valderrama EI, Giraldo C, Montoya-Lerma J, Ambrecht I, Calle Z (2006) Guía para el establecimiento y manejo de colonias artificiales de hormiga arriera *Atta cephalotes*(Hymenoptera: Myrmicinae). Boletín del Museo de Entomología de la Universidad del Valle 7(2):9-16
71. Van Soest PJ (1994) Nutritional ecology of the ruminant. Cornell University Press, Ithaca.
72. Vogt JT, Smith WA, Grantham RA, Wright RE (2003) Effects of temperature and season on Solenopsis invicta Buren (Hymenoptera: Formicidae) foraging in Oklahoma. Environ Entomol 32(3):447-451
73. Werdin González JO, Murray AP, Ferrero AA (2008) Bioactividad de aceites esenciales de*Schinus molle* var. areira (Anacardiaceae) en ninfas II de *Nezara viridula* (Hemiptera: Pentatomidae). Bol San Veg (Plagas) 34:367-375

Chapter 7

Continental-Scale Assessment of Provisioning Soil Functions in Europe

Gergely Tóth, Ciro Gardi, Katalin Bódis, Éva Ivits, Ece Aksoy, Arwyn Jones, Simon Jeffrey, Thorum Petursdottir, and Luca Montanarella

Joint Research Centre, Institute for Environment and Sustainability, Via Fermi 2749 TP280, Ispra 21027, Italy

ABSTRACT

Introduction

A framework is developed to link major soil functions to ecosystem services assessment. Provisioning soil functions—with primary linkages to ecosystem services—are evaluated on a continental scale in Europe.

Methods

We defined major provisioning soil functions combining the approaches proposed by the Millennium Ecosystem Assessment and the Thematic Strategy for Soil Protection of the European Union. Soil productivity was evaluated by three main land use types (cropland, grassland, forest) using a validated expert model called Soil Prod. Models include soil, climate and topographic criteria. Raw material provision capacity of soils was assessed on the basis of (i) organic carbon content and (ii) availability of soil materials for construction.

Results

A coherent system of soil function-based ecosystem services was compiled, taking into account major soil functions. We also produced new data on soil-based provisioning ecosystem services, including productivity and raw material availability. The attempts to cover the main human activities requiring materials of soil origin and to map the locations where those materials are available on a continental scale provide new insight to this field of research.

Conclusions

Soil-based ecosystem services can be assessed by the evaluation of soil functions which play a role in the production of these services. Quantitative analysis and comparison of the spatial distribution of the investigated soil functions were performed.

While crop productivity showed a general trend to increase in a northward and westward direction, local soil quality in most regions—except in the Mediterranean—can compensate for climatic handicaps to a great extent.

Comparison of areas with potential for providing ecosystem services by individual soil functions highlights the complexity of decision-making for resource utilization but also the possibilities for optimization and more conscious management.

INTRODUCTION

Soil plays a crucial role in terrestrial ecosystems and in maintaining life on Earth. Its functions which support ecosystem services to humans are manifold and complex.

This paper proposes a framework for the evaluation of soil functions that play a role in ecosystem services on a continental scale across Europe and offers an account of the repertoire of major soil functions and functional capacities of soils. Soil functions and associated services are discussed in the context of the European Union's Thematic Strategy for Soil Protection and are related to the concepts of the Millennium Ecosystem Assessment.

The attempt to characterize soil functions and ecosystem services for the terrestrial area of the EU in a spatially explicit manner is based on new data on provisioning soil ecosystem services, including productivity and raw material availability. Quantitative comparisons between major climatic zones of Europe were made with regards to the capacity of the soil to carry out provisioning functions.

Comparing the potential of individual soil functions across the EU highlights the complexity of decision-making dilemmas for resources utilization but also underlines the possibilities for resource use optimization and conscious management.

Soil plays a fundamental role in terrestrial ecosystems as a three-dimensional body that performs a wide range of ecological functions as a part of the services provided by ecosystems (Hannam and Boer 2004). The multitude of complex interactions which occur within the soil, including between biotic and abiotic compartments, gives rise to numerous soil functions (Blum 2005) which support ecosystem services, often viewed in terms of the services they provide to humans (De Groot et al. 2002; Fisher et al. 2009 Peccol and Movia 2012).

Soil functions are general or specific capabilities of soil to support various agricultural, environmental, landscape, and urban applications. Specific soil functions are manifold and may be grouped according to the Thematic Strategy for Soil Protection of the European Union (TSSPEU; CEC 2006) as (1) biomass production, (2) storing, filtering, and transforming nutrients and water, (3) hosting the biodiversity pool, (4) acting as a platform for most human activities, (5) providing raw materials, (6) acting as a carbon pool, and (7) storing geological and archaeological heritage. The focus of our current assessment has been on these seven major soil functions, which reflects our current scientific and policy perspectives towards Europe's soil resources (CEC 2006). Although the criticism may be valid that a scientific evaluation of soil functions as defined by a policy document is biased by current policy-driven perception, rather than being based purely on theoretical grounds (with possible windows towards applicability), it is generally accepted (FAO1976) that, in addition to studying soils' natural phenomena, soil resources should also be evaluated from socioeconomic viewpoints. This viewpoint is also an aim of the Millennium Ecosystem Assessment (MEA; Reid et al. 2005), the first comprehensive attempt to characterize the complex interactions between the various functions of ecosystems from the viewpoint of their services to humans. Therefore, a soil assessment derived from the recommendations of the TSSPEU and within the framework of soil functions is well justified for continental-scale applications in Europe.

A number of other classification and evaluation schemes for ecosystem services have emerged in recent years (Robinson et al. 2013), including the TEEB Assessment (2010; The Economics of Ecosystems and Biodiversity), which proposed a modification of the MEA scheme. Recommendations to link soil functions and ecosystem services have also been developed for local- to national-scale uses (Hay garth and Ritz 2009).

This paper aims to provide a framework to link major soil functions to ecosystem services assessment for continental-scale applications based on the concepts of the MEA and the TSSPEU. A further aim of the current paper is to provide an assessment of soil functions for provisioning services.

METHODS

Concept of the Assessment

The capacity of soil to perform any of the seven identified functions and to support their associated ecosystem services depends on its physical, biological, and chemical attributes (i.e., "internal" characteristics), while the realization of the performance is conditioned by natural (e.g., slope steepness) and/or anthropogenic factors ("external" controls). In order to provide quantitative figures on the levels of soil functions which are valid in the spatial context of the diverse biophysical regions of the European Union, evaluation of soil functions has to be extended and combined with landform and climatic information. As soil quality—the capacity of soil to perform its functions—is a time-dependent dynamic phenomenon (Karlen et al. 2001; Larson and Pierce1991), detecting temporal changes in soil functioning capacities is only possible if soil attributes are monitored. Since a comprehensive soil monitoring network and derived spatial database are lacking in Europe, the ambition to provide dynamic information on soil functions is unrealistic. However, characterizing soil functioning capacities on the basis of existing databases and matching this information with available land use and climate data were deemed feasible.

Most soil functions are interdependent; one function is not performed in isolation from another. Yet individual soil functions can be characterized by their primary linkages to ecosystem services and thus to services they provide to society, as is necessary for ecosystem services evaluation. Table 1 provides a framework proposed for continental-scale assessments. In the MEA, the process of soil formation, along with nutrient cycling and photosynthesis, was classified as an essential supporting service that facilitated other ecosystem services (i.e., provisioning, regulating, and cultural). As this study is concerned with soil functions, rather than underlying soil processes, a scheme modified from the MEA has been adopted, where soil functions are linked to ecosystem services through specific direct and indirect services provided by them.

Table 1: Soil functions that play a role in ecosystem services either directly, through their primary benefits to humans, or indirectly, by contributing to ecosystem functioning that eventually leads to benefits for humans

Soil functions (CEC2006)	**Type of linkage**	**Ecosystem services (MEA2003)**			
		Supporting	**Provisioning**	**Regulating**	**Cultural**
Biomass production	Direct		Providing food, feed, wood, fiber, biofuels		
	Indirect	Supporting ecosystem functions through primary production		Regulating carbon sequestration	Contributing to traditions, spiritual inspiration
Storing, filtering and transforming nutrients and water	Direct			Regulating water and nutrient availability	
	Indirect	Supporting ecosystem functions through water and nutrient cycling			
Hosting biodiversity	Direct	Supporting ecosystem functioning by hosting biodiversity			
	Indirect		Providing pharmaceuticals and bio chemicals through gene reserve	Regulating crop pollination, pest, and disease control through living organisms	Contributing to scientific discovery
Platform for human activities	Direct	Supporting human habitats			
	Indirect				
Source of raw materials	Direct		Providing minerals and soil organic matter		
	Indirect				

Carbon pool	Direct			Regulating atmospheric CO_2	
	Indirect	Supporting soil structure, nutrient capacity, etc.		Regulating soil system	
Storing geological and archaeological heritage	Direct				Cultural heritage values (natural sciences, history, and anthropology)
	Indirect	Geological heritage supporting the maintenance of ecosystem dynamic equilibrium			

Tóth et al.

Tóth et al. Ecological Processes 2013 2:32, doi: 10.1186/2192-1709-2-32

Soil Functions That Play a Direct Role in Provisioning Ecosystem Services

Soil functions include material outputs from soil, which can be organic or inorganic in nature. Organic materials include living aboveground biomass and subsurface organic matter. Inorganic materials include all forms of mineral compounds in the soil.

Biomass production and raw material provision are the two soil functions with material end products which are directly used by humans. However, it needs to be stressed that biomass production is a renewable function of soil, while the provision of raw material in most cases is destructive for the soil body. This distinction needs to be considered when planning for sustainable soil use.

To evaluate the performance of the biomass production function, traditional biophysical land evaluation methods can be used. For the evaluation of the raw material provision functions, the definition of the criteria for soil-originated raw material needs to be established. Firstly, it is important to set distinct criteria for soil and non-soil materials. Most mineral raw materials used by humans are of geological origin and are generally excavated by mining operations. Although soil parent material is usually considered as part of the pedon (e.g., soil taxonomy), in the raw material provisioning soil service, only soil layers above the parent material are considered. This designation also corresponds to the soil characterization provided by the TSSP, according to which "soil is generally defined as the top layer of the Earth's crust, formed by mineral particles, organic matter, water, air and living organisms" (CEC 2006).

Based on the above considerations, the biomass production function and raw material availability in the soil layers above the parent material were assessed during the evaluation of provisioning soil ecosystem services.

Biomass production of a given soil depends on the geographic location (climatic, hydrological, and terrain conditions), the land use type, land management, and the vegetation under consideration. In a continental-scale assessment, a land use–specific evaluation should be performed for the main biomass-producing land use types, namely cropland, grassland, and forest.

For continental-scale evaluations of raw materials in soils we must limit ourselves to the issues apparent over larger areas. While acknowledging the significance of locally important soil characteristics (such as high quality kaolinite for ceramics; bentonite for metal casting, well drilling, or food additives; clay for various applications), this study has focused on two applications: (1) organic soil (for horticultural and other applications) and (2) soil materials in construction. These two fields of applications involve by far the largest amounts of excavated soils in Europe. It is worth noting that a regional or local soil quality assessment might consider other elements (and with different weights). It is also worthwhile to mention that the growing demand for the services provided by the genetic pool of soil biota (e.g., from pharmaceutical industry) might lead to extraction of some soil material locally. However, the magnitudes of these kinds of extractions involve minor areas on the local scale and are not significant on a regional scale, and as such do not significantly affect the capacity of soil to provide other ecosystem services.

Attention should be called to the major difference between renewable biomass production and destructive use of soil substances. Authors strongly stress the need for considering sustainable soil use with regards to provisioning soil functions.

Databases

Soil Data

The European Soil Database (ESDB; EC 2003) was used as the soil information source in this study. Specifically, the Soil Geographical Database of Eurasia (SGDBE) and the PedoTransfer Rules Database (PTRDB) components of the ESDB were utilized.

The SGDBE, at scale 1:1,000,000, and the PTRDB are parts of the European Soil Information System (Panagos 2006; van Liedekerke et al. 2004). The SGDBE consists of both a geometric dataset and a semantic dataset (set of attribute files) which link attribute values to the polygons of the geometrical dataset. The database contains a list of soil typological units (STU). Besides the higher level soil taxonomic classification units represented by a soil name, these units are described by variables (attributes) specifying the nature and properties of the soils, for example, texture, water regime, or stoniness. As the original taxonomic information of the SGDBE is based on the 1990 FAO-UNESCO Soil Revised Legend (FAO 1990), these names and associated information content are used in the current exercise.

The PTRDB consists of a set of rules to derive new attributes with input attributes of the SGDBE (EC 2003). Pedotransfer rules related to water availability were applied in this study.

Land Use Data

The land use data of the SGDBE were exploited during the biomass productivity evaluation. The SGDBE holds information on dominant and secondary land use types of each component STU of the soil map polygons. For cropland productivity evaluation, only those STUs were considered which had cultivated land as primary or secondary land use type in the SGDBE.

The GLOBCOVER dataset developed by the European Space Agency (Bicheron et al. 2006) was used to restrict the study area separately for the main land cover classes during the validation process.

The CORINE (CO-oRdination of IN formation on the Environment; JRC-EEA 2005) land cover database for the year 2000 was used to select the extent of different land use types (cropland, grassland, forest land) to present the results.

Climate Data

Europe has diverse climatic conditions, represented by climatic zones and areas (FAO 1990; Köppen 1936). Rainfall characteristics and temperature regime seem to provide sufficient information for continental-scale agro-meteorological zonalization (Bouma 2005). The limiting effect of solar radiation is less articulated in most of Europe than that of rainfall or temperature. Differences in the radiation intensity can therefore be expressed through the above two factors and through terrain characteristics (slope and its orientation).

Soil quality evaluation for biomass productivity was therefore performed in a spatially explicit manner, taking the impact of the climatic component of soil productivity into account. Climatic zonation based on the 35 climatic areas of Hartwich et al. (2005) served as spatial units for soil productivity assessments on the continental scale. Regrouping of the climatic areas was performed to create climatic zones for soil quality assessment (Figure 1).

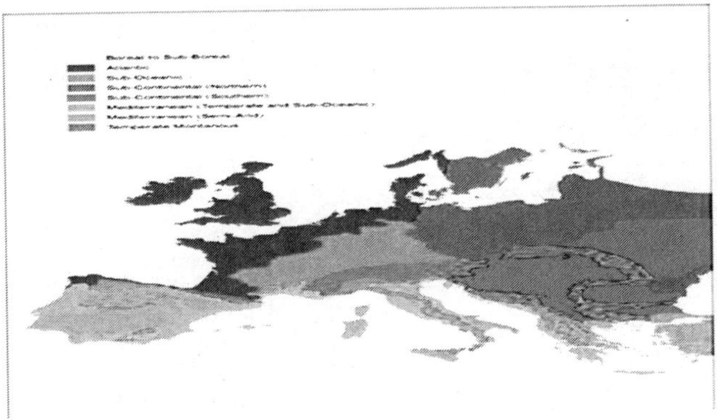

Figure 1: Climate zones of the study.

Crop and forest stand requirements regarding climatic conditions might differ to a great extent. For forest land, the length of the vegetative season and the soil water balance, calculated based on potential evapotranspiration (ETO; Penman 1948), were the aspects taken into consideration for the evaluation of forest productivity.

Topographic Data

A digital elevation model (DEM) was used to include a topographic component in the land evaluation model. The applied DEM was derived from the Shuttle Radar Topography Mission (SRTM; Rabus et al. 2003), which provided a dataset with a grid cell of 90 m.

Validation Datasets

In order to address the scientific reliability of the developed biomass productivity models, a validation procedure was applied using an independent dataset. A remote sensing–derived productivity indicator was used to validate all three land use–specific soil productivity models independently.

SPOT VEGETATION[a] decadal data were used to derive an approximation of biomass for the whole European continent. Data from the VEGETATION Program are frequently used in global and continental studies that supply input to General Circulation Models derived from measurements of the land cover and of the seasonal and long-term variations in vegetation dynamics. Studies that address the effect of biosphere processes and land cover characterization, the estimations of land cover variables as well as their dynamics, and the quantification of the mechanisms by which vegetation cover and ecosystems are interlinked all benefit from the VEGETATION Program. The SPOT data were corrected for system errors (misregistration of the different channels, calibration of all detectors along the line-array detectors for each spectral band) and resampled to geographic projections for multitemporal analysis as well as for comparison with high resolution data. For the present study, the 10-day synthesis or the maximum Normalized Difference Vegetation Index (NDVI) composite was used as is commonly accepted in the scientific community.

The NDVI time series was smoothed in order to remove short peaks and drop-offs due to noise. This pre-processing resulted in the reference time series from which the first (January) and last (December) absolute minimum values were derived for each year. The integral surface under the NDVI curve defined by the first and last minima of the year was derived and used as the approximation of net primary production (NPP; Figure 2). Ten years of NPP values were averaged for each pixel, and a final NPP dataset was produced (Figure 3). Many studies have already discussed the use of the integral surface of the annual NDVI time series curve as an approximation of aboveground biomass production (for an overview see Hill et al. 2008; Hell den and Tottrup2008). In the present study the NPP dataset was used to validate the soil productivity model in a geographically explicit manner.

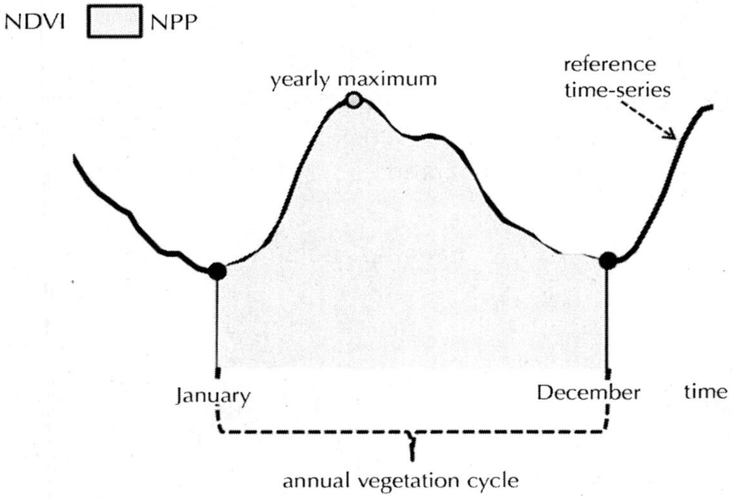

Figure 2: Schematic representation of gross annual biomass fraction (GABF) calculated from the NDVI time series for 1 year.

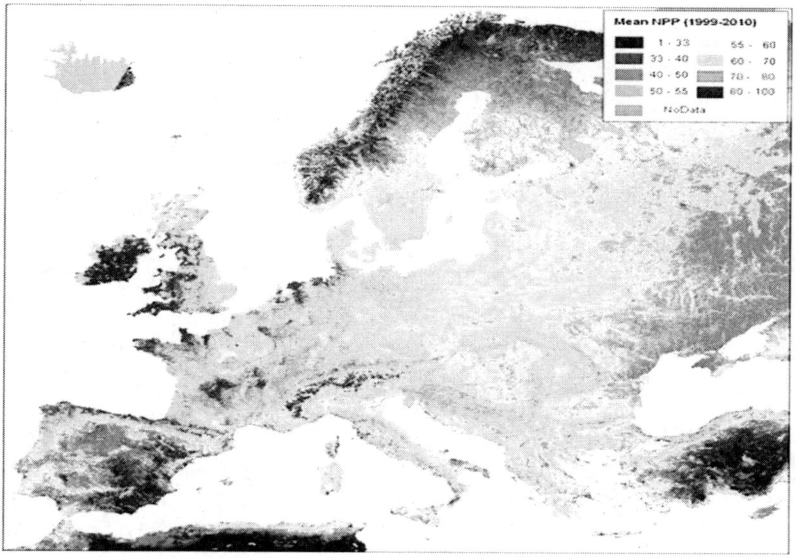

Figure 3: Mean GABF derived from time series of SPOT VEGETATION NDVI images averaged over the time series 1999–2010 for Europe.

For the raw material provisioning service, no validation datasets are available, therefore no similar validation was performed for this model.

Evaluation of Soil Provisioning Functional Capacities

Biomass Production

The degree to which the soil carries out its biomass production service was evaluated on the basis of soil properties under prevailing climatic and topographical conditions. Since productivity is a result of the interaction of soil, climatic, and topographical conditions, these factors need to be assessed in their complexity.

In addition to geophysical conditions, soil productivity also depends on the type of land use. The assessment of the European Environmental Agency (EEA 2006) shows that the three major land use types dominating the land cover of Europe are arable land with a share of 33%, pastures and mosaics with a share of 23%, and forests with a share of 29%. The aggregated share of these three types of land uses sums up to 85% of the total land and freshwater surfaces of the 24 countries of Europe assessed by the EEA (2006). Besides these major land use types, there are a number of specific regionally characterized land uses in Europe. There might also be considerable differences in the land utilization within the main land use types. However, for a continental-scale assessment of biomass productivity, the productivity patterns were evaluated according to the three major land use types. Models were therefore developed to describe general orders of soil productivity within the three land use types, namely for pasture/ grassland, cropland, and forest.

Calculations were performed in a spatially explicit manner, taking climatic and topographical conditions into account. Productivity models were built to reflect rain-fed conditions. The description of temporal variability of productivity or the estimation of provision productivity by means of actual yields was not among the aims. Results are presented in land use–specific maps (e.g., cropland productivity for areas of rain-fed arable lands, forest biomass productivity for forest lands, and grassland productivity for pastures and mosaics).

Biomass Production on Grasslands and Arable Lands

Productivity differences of similar soils under intensive rain-fed agricultural use and under grassland land cover vary with the changing availability of precipitation and differences in temperature regimes. For instance, in a temperate sub-oceanic climate, the rather stable thermal regime and balanced water availability of medium to high amounts of precipitation not only secure plant-available water on most soil types throughout the growing period but facilitate decomposition and weathering throughout the year. These processes are limited under temperate continental and Mediterranean climates due to cold and/or arid periods in most years. With increasing aridity in prevailing climates, the importance of soil physical and chemical properties in water and nutrient supply to plants gains increasing significance. Based on this principle of soil productivity processes, ranking of inherent productivity of soil was performed separately for the major climatic areas. While fertility is defined as the ability of the soil to provide nutrients and water, productivity refers to the capacity to supply nutrients and water and thus produce plant biomass at a given level/quantity. Inherent productivity in this context means soil productivity before human interference. In this study, inherent soil fertility—in this context productivity—is considered as a proxy for grassland productivity. Although the existence of different nutrient inputs on grasslands in different regions of the European Union is recognized, due to the lack of adequate data, grassland fertilization and other management factors were not taken into account. Cropland soil productivity was evaluated by the extension of inherent productivity with a management factor, as described below.

Grassland/Pasture

In the first step of the evaluation process, eight characteristic European climate systems were identified where the complex effects of water availability and thermal regime are distinct for soil processes and plant growth. The 35 climatic areas of Europe (Hartwich et al. 2005) were arranged into eight climatic groups accordingly (Figure 1). The climatic groups comprise regions where the following concepts of soil productivity processes prevail: boreal to sub-boreal (CZ1), Atlantic (CZ2), sub-oceanic (CZ3), northern sub-continental (CZ4), southern sub-continental, (CZ5), Mediterranean semi-arid (CZ6), and Mediterranean (temperate and sub-oceanic) (CZ7), and temperate mountainous (CZ8). Long-term average inherent soil productivity ratings were assigned to soils of each region in a spatially explicit manner. Temporally variably climatic hazards of productivity (frost, water logging, etc.) were not taken into account in the current study.

Inherent soil productivity estimates were derived from the original taxonomic component (second-level taxonomic soil units) and soil attribute information of the soil database. Second-level taxonomic soil units were grouped into five inherent capability classes according to their relative productivity in each climatic zone. CZ1 includes 59 different soil units, CZ2 has 113, CZ3 has 122, CZ4 has 117, CZ5 has 128, CZ6 has 96, CZ7 has 103, and CZ8 has 131. This approach allows productivity assessment of soil types at different climatic regions. Table 2 provides an example of the classification.

Table 2: Examples of climate-zone-based productivity classification reference (10 soil units out of the total 128 present in the southern sub-continental climate zone are displayed here)

Inherent capability classes (southern sub-continental climate)				
V.	**IV.**	**III.**	**II.**	**I.**
Calcaric Lithosol	Orthic Rendzina	Gleyo-Dystric Luvisol	Albic Luvisol	Haplic Phaeozem
Gleyic Solonchak	Gleyic Podzoluvisol	Dystric Regosol	Chromic Cambisol	Luvic Chernozem

Tóth et al.

Tóth et al. Ecological Processes 2013 2:32, doi: 10.1186/2192-1709-2-32

In parallel, STUs were also rated according to the available water capacities (AWC) of topsoil and subsoil. According to the pedotransfer rules of the PTRDB (EC 2003) and based on physical soil properties and soil depth, AWCs of both the topsoil and the subsoil were calculated. Soils were grouped into four classes on the basis of the water-storing capacity of the profile (Table 3).

Table 3: Examples of water-storing capacity classifications

Water-storing capacity classes	AWC of topsoil	AWC of subsoil	Depth to impermeable layer
I.	Very high	Very high	Deep
	High	Very high	Deep
II.	Medium	High	Deep
	High	Medium	Deep
III.	Medium	Low	Deep
	High	Very low	Deep
IV.	High	Very low	Shallow
	Medium	Low	Shallow

Tóth et al.

Tóth et al. Ecological Processes 2013 2:32, doi: 10.1186/2192-1709-2-32

In the next step of the productivity assessment, an evaluation matrix was set up for the eight climatic regions and the five inherent capability classes. Productivity scores between 1 and 8 were assigned for each cell in the matrix, based on the complex evaluation of the climate-dependent relative inherent fertility of soils (Table 4).

Table 4: Productivity scores for the soil capability classes by climatic zones

CZ1	CZ2	CZ3	CZ4	CZ5	CZ6	CZ7	CZ8	Score
								8
I.	II.	I.	I.				I.	7
II.	III.	II.	II.		I.	I.	II.	6
III.	IV.	III.	III.	I.	II.	II.	III.	5
IV.	V.	IV.	IV.	II.	III.	III.	IV.	4
V.		V.	V.	III.	IV.	IV.	V.	3
				IV.	V.	V.		2
				V.				1

Tóth et al.

Tóth et al. Ecological Processes 2013 2:32, doi: 10.1186/2192-1709-2-32

Water-storing capacity classes (Table 3) were assigned a multiplication factor between 0.75 and 1 for each climate zone. These factors were used to multiply the productivity scores of each soil unit of the climate zones.

The position of soil units in the capability classification matrix, productivity scores of capability classes by climate zones, and multiplication factors to account for the water-storing capacities by climate zones were first defined by expert judgment and then refined and completed through an iterative analytical process with a series of cross validations using regression analysis (GWR) with the validation dataset as dependent variable (see below for details on the validation process).

The result of this procedure is the soil productivity index for each soil typological unit in the climate zones. The index score 1 represents the poorest and 8 the highest productivity soil. The corresponding inherent productivity scores were assigned to each STU in the SGDBE. A spatially weighted average of inherent productivity score was calculated for each soil mapping unit on the basis of the proportional areal shares of the STUs within the mapping units.

Correction measures based on topographic conditions were finally applied to arrive at the final evaluation scores of inherent soil productivity. Correction coefficients of slope and aspects applied according to the D-e-Meter land evaluation system (Tóth 2009) were adapted for this continental-scale study.

Croplands

To evaluate the biomass productivity of soils under arable cultivation, the model was extended with a new module. Since the productivity of soil is due to its inherent fertility and to the effect of management—mainly nutrient input—the effect of fertilization was considered in this module. While acknowledging the importance of the applied technology of soil use on the actual productivity of soil, detailed distinctions of management practices were not considered in this study. The goal of this study was solely to determine soil productivity, i.e., the capacity of soil to supply nutrients, water, and rooting medium for plants, in a comparative manner and not to assess the effects of management differences other than nutrient supply. We also assume that in the case of the realization of biomass productive capacity, technological advancement plays a relatively insignificant role within distinct regions in European Union. However, in certain regions, where complex socioeconomic and biophysical conditions diversify the technological levels, this statement might not hold. The efficiency of input use, i.e., the selection of the most appropriate techniques and the amount of input, is largely determined by biophysical conditions (e.g., large regional crop yield differences within France (Eurostat 2013), are due to biophysical differences rather than differences in available technology, capital, or other socio-economic conditions). However, the technological advancement and input intensity of agriculture across regions in the European Union varies, and this variation contributes to differences in crop yields.

It was not the goal of this study to evaluate all management-related yield responses of soils, but rather, to evaluate the effect of management, we considered the influence of fertilization. To do this, second-level taxonomic soil units were grouped into five classes in each climatic zone, according to the magnitude of their expected productivity increase due to fertilization (Table 5).

Table 5: Sample reference table for classifying the response of different soil types to fertilization on a climate zone basis (10 of 128 soil units present in the southern sub-continental climate zone are shown here)

Fertilizer response classes (southern sub-continental climate)				
V.	**IV.**	**III.**	**II.**	**I.**
Luvic Ranker	Eutric Gleysol	Calcaric Phaeozem	Albic Luvisol	Dystric Fluvisol
Cambic Rendzina	Chromic Vertisol	Dystric Podzoluvisol	Calcic Cambisol	Calcaric Cambisol

Tóth et al.

Tóth et al. Ecological Processes 2013 2:32, doi: 10.1186/2192-1709-2-32

Then, a fertilizer response score was assigned to each soil unit in the eight climatic zones. Soils with the largest relative fertility increase received the maximum of 8 points and soils with little influence of fertilization received 1 point. For example, an Albic Arenosol in the sub-oceanic climate has among the highest relative productivity increases when properly fertilized, while fertilization has little effect on the crop productivity of Calcaric Rendzinas in the Mediterranean.

To calculate soil productivity for the cropland land use type, the inherent soil productivity and the fertilizer response scores were aggregated, assigning a mechanical weight to the fertilizer response indices. This weight resulted in the best model fit at the end of an iterative statistical validation process (see below for details of the statistical validation). Spatially weighted averages of productivity scores were calculated for the SMUs of the SGDBE. In order to avoid the bias originating from the evaluation of non-cropland soils, only those STUs were considered which had cultivated land as the primary or secondary land use type in the SGDBE. Finally, similarly to the concluding step of the grassland productivity evaluation, correction coefficients were applied to evaluate the effect of the topography (slope and aspect) on the productivity of cropland soils.

Biomass Production on Forest Lands

The productivity of a forest, in terms of biomass, is the result of the interaction of tree species, soil, and climate. Ecosystem research has provided sufficient information about key processes such as photosynthesis, transpiration, and decomposition to allow the construction of simulation models that predict properties such as biomass production, soil carbon storage, and nitrogen cycling rates (e.g., Aber et al. 1997; Parton et al. 1988; Raich et al. 1991; Running and Gower1991).

Several of these models are based on complex input data sets, consisting of a series of climate, soil, and forest stand data (i.e., maximum and minimum daily temperature, vapor pressure, solar radiation, total monthly precipitation, forest type, and plant-available soil water holding capacity).

The aim of this study, however, was to predict, in general terms, the forest productivity potential of European soils. For this purpose, two main processes related to forest productivity were evaluated: the length of the vegetative season and the soil water balance. The concept behind this simplistic approach is that the net ecosystem productivity (NEP) of a forest can be limited by two main factors: the air temperature, which determines the length of the vegetative season, and the amount of water from soils that the plants can evapo-transpire. Accordingly the northern latitude forests will be limited by the relatively short vegetative season, while the Mediterranean forests will be limited by the relatively low water availability, which in turn is related to both climate and soil characteristics.

The equation used for the calculation of forest productivity is shown below:

$$\text{Forest productivity} = (SWBN_x / SWBN_{max}) \times \text{vegetative length}$$

Where SWBNx is the soil water balance normalized at a given location (mm), SWBNmax is the maximum soil water balance (mm), and vegetative length is the length of the vegetative season (days). The soil water balance is calculated as:

$$\Sigma(P_i - ETP_i) + AWC(\text{mm})$$

Where Pi is monthly precipitation, ETPi is monthly potential evapotranspiration (Kc =1), and AWC is the available water content (mm).

The values obtained, ranging from 0 to 313, were grouped into 10 equal classes. The output map was masked based on the actual forest distribution as defined by the CORINE Land Cover database.

Results of all three productivity evaluations (grassland, cropland, forest) were scaled to index scores ranging from 1 to 10 showing the relative fertility of soils expressed in relative index values without units.

Provisioning Raw Materials

The evaluation of raw material availability from soil origin is limited to resources above the parent material. The assessment included the following two options, with the relevant criteria:

- Peat (for horticultural and other applications) and organic topsoil: All organic soils (Histosols) are considered and no mineral soils are considered.
- Soil materials for construction: To assess the quality of soils to provide construction materials, the approach presented in the Soil Atlas of Europe [b] was applied, and the presence of sand and gravel was examined for this function. Criteria:
- Coarse texture (clay < 18% and sand > 65%) and/or
- Stones and gravel content are dominant in the horizon (> 80% by volume)

In addition to the above construction materials, loamy clay is often used to produce bricks and tiles. Although the Soil Atlas of Europe does not consider these as among the dominant soil-based construction materials, their widespread usage could justify the inclusion of loamy-clay (sub) soils in the analysis. However, the currently available continental soil databases contain no information on soil texture at this level of detail; therefore such an analysis was not feasible at this time.

[b]The Soil Atlas of Europe states: "Sand and gravel are the most accessible basic raw materials for the construction industry and are used as concrete aggregates, road base, mixed with bitumen for road surfaces, construction fill, snow and ice control (gritting), railroad ballast, roofing granules and water filtration systems."

The evaluation did not account for any economic considerations (i.e., only the availability of raw material was assessed). Results are presented as the proportional availability of these materials in the mapping units. For Cyprus, no information on texture and stone and gravel content was available and, therefore Cyprus was excluded from this analysis.

Statistical Validation

The geographically weighted regression (GWR) method (Brunsdon et al. 1996; Fotheringham et al.2002) was selected to validate the model results against the measured productivity indicators. The GWR method was selected as it allows for the possibility of assessing relationships in spatially explicit information. Spatial data are rarely consistent due to random sampling variations, the different relationships between variables across space, or model misspecifications of reality (Fotheringham et al. 2002). However, the application of standard regression methods to spatial data would result in the assumption of the spatial phenomenon being constant over space.

GWR is a statistical technique that allows the modeling of processes that vary spatially. Consider the ordinary least squares (OLS) regression model given by

$$y_i = a_0 + \sum_k a_k x_{ik} + e_i,$$

Where *ei* is the error term, *a* represents the vector of global parameters to be estimated, *x* is a matrix of independent variables and *y* represents a vector of observations on the dependent variable.

GWR extends the traditional regression framework by allowing local rather than global parameters to be estimated. Thus the OLS regressions model is rewritten as:

$$y_i = a_0(u_i, v_i) + \sum_k a_k(u_i, v_i) x_{ik} + e_i,$$

where (*ui,vi*) denotes the coordinates of the *i*th point in space and *ak*(*ui,vi*) is a realization of the continuous function *ak*(*u,v*) at point *i* (Brunsdon et al. 1996).

This extension allows measurements of a continuous surface of parameters to be taken and to denote the spatial variability of the surface because observed data near point *i* have more of an influence in the estimation than do data located farther from *i*. Thus GWR becomes a weighted least squares regression where observations are weighted in accordance with the proximity to point *i* and the weight decreases with increasing distance from with *i*.

The biomass productivity spatial datasets were divided—both from our land evaluation exercise and from the satellite observations—by land use, based on small regions for special diagnoses (NUTS3; Eurostat 2011). Analyses of statistical correlation between the datasets were performed using the GWR method as described above.

Per pixel biomass values estimated from the land evaluation exercise and from the satellite-based observations were averaged within the regions of the NUTS3 dataset (Eurostat 2011) for the countries of the European Union. The GWR analysis was carried out separately for croplands, grasslands and forests with the remote sensing–derived biomass estimates as the dependent variable and the land cover–specific biomass estimates as independent variables. Each regression run was weighted by the number of pixels within the NUTS3 polygons.

The result of the GWR model for croplands is valid with a goodness of fit characterized by a coefficient of determination (adjusted r^2) of 0.73. The adjusted r^2 figure for grassland is 0.85 and for forests 0.90. These results can be considered as a confirmation of the validity of the soil productivity evaluation approach.

RESULTS AND DISCUSSIONS

Grassland Soil Productivity Map of the EU

The application of a geographically explicit spatial soil grassland productivity model showed high consistency with the data obtained from remote sensing ($R^2 = 0.85$).

Based on the cumulative relative productivity indices, the soils in the Atlantic climatic zone have the highest potential for grassland production. They contribute nearly half of the potential for the European Union (44.6%), although only 38.4% of Europe's grassland areas are found in the Atlantic zone (Table 6). The zone productivity factor of grassland soils in this climate zone is 1.2, indicating 20% higher relative productivity indices on average when compared to the mean of the EU. In contrast, grasslands of the semi-arid Mediterranean region have relative productivity indices that are 40% lower than the average of the EU. Apart from the northern sub-continental zone

(zone productivity factor = 1), all other zones have relative productivity indices for grasslands that are lower than the average of the EU. This fact indicates that, in most of the climatic zones (CZ1, CZ3, CZ5, CZ6, CZ7, CZ8) of Europe, grassland ecosystems are on soils of lower production potentials.

Table 6: Summary statistics of productivity indices for soils in the European Union by major land use classes and main climatic zones

Climate zone [a]	Area		Productivity index				Zone productivity factor [c]
	km²	**Percentage of EU**	**Mean**	**Standard dev**	**Sum**	**Percentage of total [b]**	
Grassland							
1	21,571	4.5	5.8	1.7	124,963	4.3	0.9
2	184,363	38.4	7.1	2.1	1,305,600	44.6	1.2
3	88,910	18.5	5.8	2.0	515,760	17.6	0.9
4	55,982	11.7	6.3	1.3	350,335	12.0	1.0
5	33,309	6.9	5.8	1.2	192,966	6.6	0.9
6	21,959	4.6	3.8	1.2	84,031	2.9	0.6
7	16,170	3.4	4.2	1.3	68,612	2.3	0.7
8	57,306	11.9	5.0	1.4	286,378	9.8	0.8
Cropland							
1	32,458	2.4	6.1	1.0	198,645	2.5	1.0
2	308,693	23.0	7.1	1.3	2,189,030	27.6	1.2
3	170,655	12.7	5.7	1.6	974,337	12.3	1.0
4	337,121	25.1	6.2	1.1	2,099,960	26.5	1.1
5	175,914	13.1	5.8	1.0	1,013,880	12.8	1.0
6	174,171	13.0	4.0	1.2	698,184	8.8	0.7
7	109,687	8.2	5.0	1.2	545,528	6.9	0.8
8	35,101	2.6	5.7	1.3	198,537	2.5	1.0
Forest							

1	531,907	30.9	2.3	0.8	1,202,780	19.2	0.6
2	120,702	7.0	5.0	1.7	606,879	9.7	1.4
3	233,015	13.5	4.8	1.2	1,111,580	17.8	1.3
4	276,591	16.1	3.3	0.8	923,848	14.8	0.9
5	86,110	5.0	4.5	1.1	385,547	6.2	1.2
6	113,045	6.6	4.1	1.1	460,481	7.4	1.1
7	151,472	8.8	4.8	1.5	731,081	11.7	1.3
8	207,675	12.1	4.0	1.2	836,053	13.4	1.1

[a]1: Boreal to sub-boreal, 2: Atlantic, 3: sub-oceanic, 4: sub-continental (northern), 5: sub-continental (southern), 6: Mediterranean (semi-arid), 7: Mediterranean (temperate and sub-oceanic), 8: temperate mountainous. [b] Expressed in terms of productivity index. [c]Zone productivity factor = % of total productivity indices/% of total area in the EU (h = g/b).

Tóth et al.

Tóth et al. Ecological Processes 2013 2:32, doi: 10.1186/2192-1709-2-32

The dependency of productivity on climate was statistically significant at the continental scale. Up to an estimated average of 60% of the grassland productivity differences between the most favorable (Atlantic) and least favorable (Mediterranean semi-arid) regions of the EU can be attributed to climatic factors, as indicated by the differences in mean productivity indices between the climatic zones and standard deviations within the climatic zones (Table 6). The influence of regional soil variability expressed as the standard deviations of the soil productivity indices might exceed the figure attributed to the influence of climate between different climate zones. These results confirm the general understanding of the importance of water availability in soil and the synergism between optimum temperature, water, and nutrient regimes and also present these interdependencies in a quantified manner in a spatial context.

Results of the inherent soil productivity evaluation are shown in the grassland productivity map of the European Union (Figure 4).

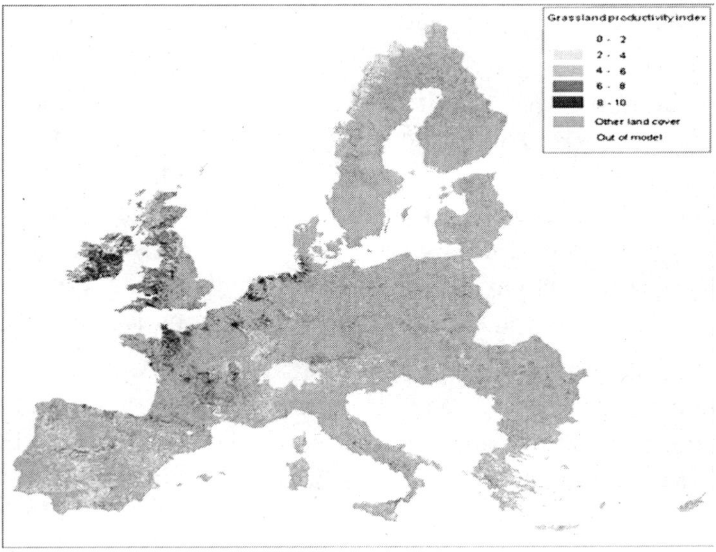

Figure 4: Soil biomass productivity of grasslands and pastures of the European Union.

Cropland Soil Productivity Map of the EU

A comparative analysis between modeled and remote sensing–derived soil productivity indicators demonstrated a significant climate dependency in the general pattern of cropland productivity of soils in the EU. However, a strong influence of soil type and soil properties was also verified.

Productivity showed a general trend of increasing in a northward and westward direction, however, local soil quality in most regions—except in the Mediterranean—can compensate for the climatic "handicap" to a great extent. While soils of the Atlantic region (CZ2), as was the case of the grassland ecosystem, have the highest relative potential to perform biomass provisioning ecosystem services in the cropland agro-ecosystem, this relative advantage is not as pronounced as in the case of grasslands (Table 6). In fact, soils of the northern sub-continental zone (CZ4; with 25.1% of the croplands of the EU) also perform above the EU average (zone productivity factor = 1.1), and another four zones (CZ1, CZ3, CZ5, CZ8), with a total share of 31% from the croplands of the EU, have, in general, soil productivity near the average of the continent (zone productivity factor = 1; Table 6). Among these four zones, the sub-oceanic zone (CZ3) possesses the highest intra-variability, as indicated by the standard deviation of the productivity assessment. Cropland soils of the Mediterranean (CZ7 and CZ8) with a combined area share of 21.2% have substantially lower rain-fed crop productivity compared to the EU's average, demonstrated by zone productivity factors of 0.8 and 0.7, respectively.

It is worth noting that irrigation can significantly increase the productivity of soils in the Mediterranean and to some extent also in the sub-continental region, where water is the limiting factor for crop growth. However, to assess (actual and potential) irrigated productivity of soil was out of the scope of this study. Nevertheless, future research should be carried out to analyze this aspect as well.

Results of soil productivity evaluation for croplands of the European Union are presented in Figure 5.

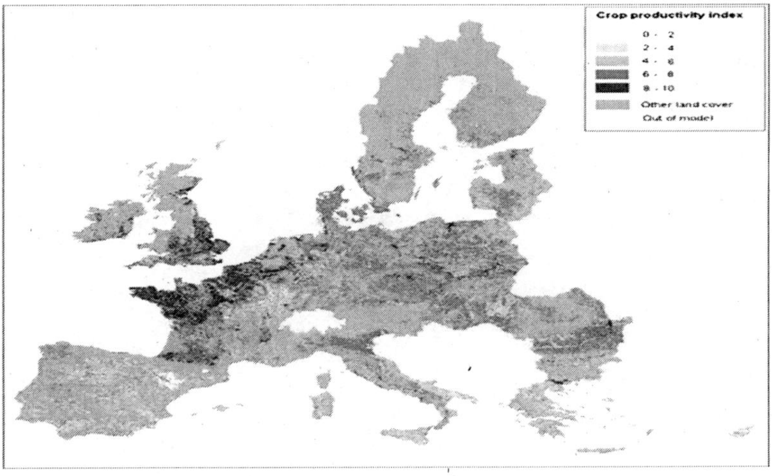

Figure 5: Soil biomass productivity of croplands in the European Union.

Forest Soil Productivity Map of the EU

Forest productivity is essentially the expression of the interaction between climate and soil factors. The productivity of forest soils is a function of their ability to hold and make water available to trees (available water capacity, AWC). Their nutrient status is also important. Available water capacity is a function of the rooting depth and texture of soils.

A common measure, especially in United States, for the evaluation of the collective influence of soil factors on forest growth is the site index (SI). The most important factors determining the SI are the topsoil depth and soil texture. The AWC parameter is strongly correlated with soil depth and soil texture and consequently can be used as a proxy of SI.

From the data shown in Figure 6, the climate driver is the most evident at larger scales, while the influence of soil can be detected mainly at more detailed scales. In fact, zone productivity factors (Table 6) have the widest range across the climate zones in Europe for forest land among the three land use classes assessed. The zone productivity factor is 1.4 for the Atlantic zone (CZ2), but only 0.6 for sub-boreal to boreal climates. In general terms, it is clear that climate is responsible for forest productivity variation at the regional scale while soil characteristics cause local (i.e., at scales of 10–100 km^2) variations.

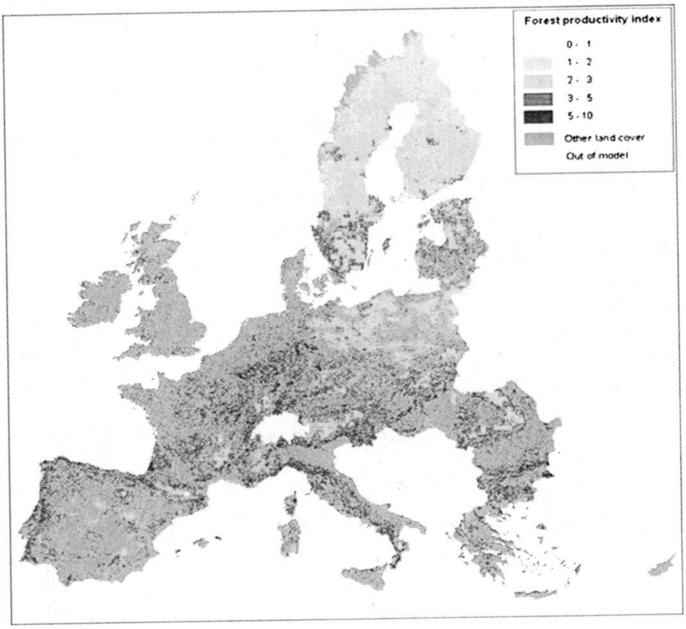

Figure 6: Soil biomass productivity of forest areas in the European Union.

It is evident that in the northern areas the favorable moisture conditions are negated to some extent by the shorter length of the vegetative season, while in the Mediterranean climate the length of the vegetative season is not a constraint and water availability during summer represents the most important limiting factor.

The soil parameter taken into consideration in the proposed approach was the AWC, for which only three classes are available within the European Soil Geographic Database: 120, 165, and 220 mm. The effect of soil is evident at a local scale, where it can result in a large variation in the absolute values of soil productivity.

Continental-Scale Map of Soil-Extractable Raw Materials of the EU

Raw material provisioning of the mapping units was calculated on the basis of the proportional shares of the STUs with raw material content in the area of each mapping unit. Once again it is important to stress that geological maps showing availability of materials worthwhile for human use that are present below the soil cover were not considered in this assessment and might show a very different pattern. The result of the assessment is presented in Figure 7.

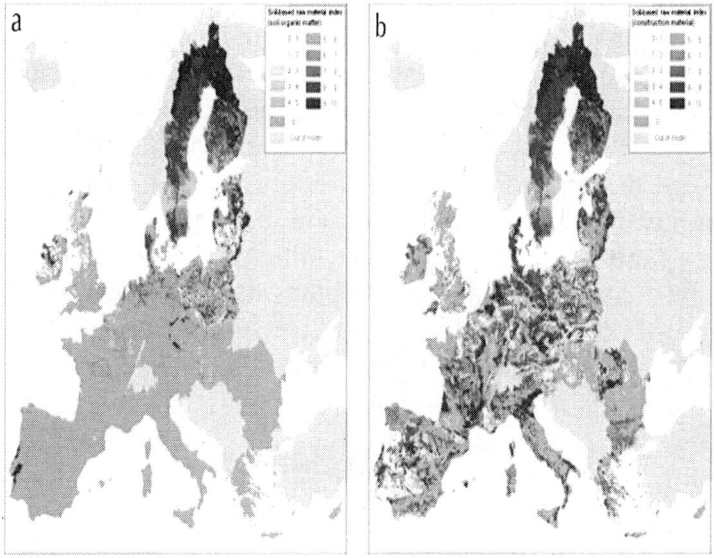

Figure 7: Raw material availability from soils of the European Union. a) Organic soil material; b) soil material for constructions.

It is well acknowledged that this first approximation to highlight the availability of soil-born raw materials on the continental scale might be biased (1) by the classification of soil materials of human interest for excavation and (2) by the thematic and geographic limitations of the dataset. However, the attempt to consider the main human activities that require materials of soil origin and to map the locations where those materials are available on a continental scale provides new insight to this field of research.

These maps showing a continental overview can suggest opportunities for raw material extraction from soil which can serve current or future needs for a number of applications, such as construction and the health industry (soil organic matter). The maps show that northern and north-eastern Europe have large stocks for both applications, while most other regions of the continent do not have large reserves of soil organic matter but are generally well suited for extraction of construction materials, with substantial intraregional differences.

The comparison of maps showing the raw material availability of European soils (Figure 7) and productivity of various land uses (Figures 4, 5, and 6) also proves the need for separate assessment of the components of soil quality. In general terms the geographic locations of the most suitable zones for agriculture and for procuring soil-borne materials are different. On the other hand, there are more overlaps between prime areas of forest lands and zones with potential to provide raw materials from soil. While the first assumption is positive for resource use planning, the second highlights the dilemma that land use planners usually face: the utilization of one service of the land might negatively influence another utilization option. With the evaluation of other soil functions and ecosystem services, the level of complexity for decision-making increases.

CONCLUSIONS

A coherent view of soils requires an expansion of the conventional thinking about the role of soils and consideration of soils in a multidimensional perspective in the context of sustainable development including the utilization of soil-based ecosystem services. Conscious utilization of soil ecosystem services can only be achieved on the basis of a proper accounting.

In this study, an assessment of soil functions in relation to ecosystem services was carried out for a continental-scale overview of the EU. A coherent system of soil function-based ecosystem services was devised, taking into account the major soil functions as identified in the Soil Thematic Strategy.

In this attempt to characterize soil ecosystem services for the EU in a spatially explicit manner, new data on soil-based provisioning ecosystem services were produced, including productivity and raw material availability. Results show that up to an average of 60% of grassland productivity differences can be attributed to climatic factors between the most favorable (Atlantic) and least favorable (semi-arid Mediterranean) regions of the EU. While crop productivity shows a general trend toward increasing in a northward and westward direction, local soil quality in most regions—except in the Mediterranean—can compensate for climatic handicaps to a great extent. The attempts to address the main human activities requiring materials of soil origin and to map the locations where those materials are available on a continental scale provide new insight to this field of research.

The comparison of areas with potential for providing ecosystem services by individual soil functions highlights the complexity of decision-making for resource utilization but also the possibilities for optimization and more conscious management.

With the evolution of time-series soil data based on monitoring, the increase in spatial precision of soil information, and the introduction of dynamic components to the models, comparative assessment of soil functions can be a valuable tool for decision-making in the future across geographical scales and for different stakeholder's needs.

AUTHORS' CONTRIBUTIONS

GT, CG, AJ, SJ, LM, and TP participated in the concept development. GT, CG, and AJ designed the models. KB and EA contributed to the GIS analysis. GT and ÉI designed and implemented the model validation. GT, AJ, CG, and SJ drafted the manuscript. All authors read and approved the final manuscript.

REFERENCES

1. Aber JD, Ollinger SV, Driscoll CT (1997) Modelling nitrogen saturation in forest ecosystems in response to land use and atmospheric deposition. Ecol Model 101:61-78
2. Bicheron P, Leroy M, Brockmann C, Krämer U, Miras B, Huc M, et al. (2006) GLOBCOVER: a 300m global land cover product for 2005 using ENVISAT/MERIS time series. Valencia.

3. Blum WEH (2005) Functions of soil for society and the environment. Rev Env Sci Biotechnol 4(3):p75-p79
4. Bouma E (2005) Development of comparable agro-climatic zones for the international exchange of data on the efficacy and crop safety of plant protection products. EPPO Bull 35:233-238
5. Brunsdon C, Fotheringham AS, Charlton M (1996) geographically weighted regression: a method for exploring spatial nonstationarity. Geogr Anal 28:281-289
6. CEC (Commission of the European Communities) (2006) Communication from the Commission to the Council, the European Parliament, the European Economic and Social Committee and the Committee of the Regions—Thematic strategy for soil protection, Commission of the European Communities. Brussels: COM. p 231
7. De Groot RS, Wilson MA, Boumans RMJ (2002) a typology for the classification, description and valuation of ecosystem functions, goods and services. Ecol Econ 41:393-408
8. EC (2003) European soil database (distribution version v2.0). Ispra, Italy: European Commission Joint Research Centre.
9. EEA (2006) Land accounts for Europe 1990–2000. Towards integrated land and ecosystem accounting. EEA Report No 11/2006. Copenhagen: EEA European Environmental Agency.
10. Eurostat (2011) NUTS—Nomenclature of territorial units for statistics. http://epp.eurostat.ec.europa.eu/portal/page/portal/nuts_nomenclature/introduction
11. Eurostat (2013) Agricultural statistics - Crops products. http://epp.eurostat.ec.europa.eu/portal/page/portal/agriculture/introduction
12. FAO (1976) A framework for land evaluation. FAO Soils Bulletin 32, Rome.
13. FAO/UNESCO/ISRIC (1990) revised legend of the soil map of the world. World soil resources report. Rome: FAO.
14. Fisher B, Turner RK, Morling P (2009) defining and classifying ecosystem services for decision making. Ecol Econ 68:643-653
15. Fotheringham AS, Brunsdon C, Charlton M (2002) geographically weighted regression: the analysis of spatially varying relationships. Chichester: Wiley.

16. Hannam I, Boer B (2004) Drafting legislation for sustainable soils: a guide. Gland: IUCN.

17. Hartwich R, Baritz R, Fuchs M, Krug D, Thiele S (2005) Erläuterungen zur Bodenregionenkarte der Europäischen Union and ihrer Nachbarstaaten 1:5,000,000 (version 2.0). Hannover: Bundesanstalt für Geowissenschaften und Rohstoffe (BGR).

18. Haygarth PM, Ritz K (2009) the future of soils and land use in the UK: Soil systems for the provision of land-based ecosystem services. Land Use Pol 26(suppl1):S187-S197

19. Hellden U, Tottrup C (2008) Regional desertification: a global synthesis. Glob Planet Change 64:169-176

20. JRC-EEA (2005) CORINE land cover updating for the year 2000: image 2000 and CLC2000. Products and methods. Report EUR 21757 EN. Ispra: JRC.

21. Karlen DL, Andrews SS, Doran JW (2001) Soil quality: current concepts and applications. In: Sparks DL (ed) Advances in agronomy, 74, San Diego: Academic. pp 1-40

22. Köppen W (1936) Das geographische System der Klimate. In: Koppen W, Geiger R (eds) Handbuch der Klimatologie, Berlin: IC.

23. Larson WE, Pierce FJ (1991) Conservation and enhancement of soil quality. In: Evaluation for Sustainable Land Management in the Developing World, Vol. 2: Technical papers. Bangkok, Thailand: International Board for Research and Management. IBSRAM Proceedings No. 12 (2). pp 75-203

24. MEA (Millennium Ecosystem Assessment) (2003) Ecosystems and human well-being: a framework for assessment. A report of the conceptual framework working group of the Millennium Ecosystem Assessment. Washington DC: Island Press.

25. Panagos P, Van Liedekerke M, Filippi N, Montanarella L (2006) MEUSIS: towards a new multi-scale European soil information system. ECONGEO, 5th European Congress on Regional Geoscientific Cartography and Information Systems. Barcelona (Spain). pp 175-177

26. Parton WJ, Stewart JWB, Cole CV (1988) Dynamics of C, N, P and S in grassland soils: a model. Biogeochem 5:109-131

27. Peccol E, Movia A (2012) Evaluating land consumption and soil functions to inform spatial planning. 3rd International Congress on Degrowth for Ecological Sustainability and Social Equity. Venice (Italy). pp 19-23

28. Penman HL (1948) Natural evaporation from open water, bare soil and grass. Proc R Soc Series A 193:120-146

29. Rabus B, Eineder M, Roth A, Bamler R (2003) the shuttle radar topography mission—a new class of digital elevation models acquired by spaceborne radar. Photogramm Rem Sens 57:241-262

30. Raich JW, Rastetter EB, Melillo JM, Kicklighter DW, Steudler PA, Peterson BJ, et al. (1991) Potential net primary productivity in South America: application of a global model. Ecol Appl 1:399-429

31. Reid WV, Mooney HA, Cropper A, Capistrano D, Carpenter SR, Chopra K, et al. (2005) Millenium Ecosystem Assessment synthesis report. Washington DC: Island Press.

32. Robinson DA, Hockley N, Cooper DM, Emmett BA, Keith AM, Lebron I, Reynolds B, Tipping E, Tye AM, Watts CW, Whalley WR, Black HIJ, Warren GP, Robinson JS (2013) Natural capital and ecosystem services, developing an appropriate soils framework as a basis for valuation. Soil Biol Biochem 57:1023-1033

33. Running SW, Gower S (1991) FOREST-BGC, a general model of forest ecosystem processes for regional applications. II. Dynamic carbon allocation and nitrogen budgets. Tree Physiol 9:147-160PubMed Abstract |

34. TEEB (2010) the Economics of Ecosystems and Biodiversity Ecological and Economic Foundations. In: Pushpam K (ed) Earthscan, London and Washington Chapter 5: the economics of valuing ecosystem services and biodiversity, http://www.teebweb.org/wp-content/uploads/2013/04/D0-Chapter-5-The-economics-of-valuing-ecosystem-services-and-biodiversity.pdf (accessed on 14/11/2013)

35. Tóth G (2009) Land evaluation with the D-e-Meter system. Agrochem Soil Sci 58(2):227-242

36. Van Liedekerke M, Panagos P, Montanarella L, Filippi N (2004) towards a multi-scale european soil information system. Porto: 13th EC-GIS Symposium.

Chapter 8

Characterization of Soil Organic Matter in Perhumid Natural Cypress Forest: Comparison of Humification in Different Particle-Size Fractions

Jenn-Shing Chen[1], Tay-Lung Chung[2], Guanglong Tian[3], and Chih-Yu Chiu[2]

[1]Yung-Ta Institute of Technology & Commerce, Linlo, Pingtung 90942, Taiwan

[2]Biodiversity Research Center, Academic Sinica, Nankang, Taipei 11529, Taiwan

[3]Environmental Monitoring and Research Division, Monitoring and Research Department, Metropolitan Water Reclamation District of Greater Chicago (MWRD), Lue-Hing R&D Laboratory, 6001 W. Pershing Road, Cicero, IL 60804, USA

ABSTRACT

Background

The *Chamaecyparis* forest is a valuable natural resource in eastern Asia. The characteristics of soil humic substances and the influence of environmental factors in natural *Chamaecyparis* forests in subtropical mountain regions are poorly understood. The study site of a perhumid *Chamaecyparis*forest is in the Yuanyang Lake Preserved Area in northcentral Taiwan. We collected samples from organic horizons (Oi, Oe and Oa) and from the surface horizon (O/A horizon) at the summit, footslope and lakeshore to characterize the composition of the soil organic matter. Samples of organic horizons were dried and ground, and those of the O/A horizon were passed through wet sieving for different particle-size fractions before analysis. The C chemical structure in the samples was determined with CP/MAS ^{13}C NMR spectra.

Results

The ratios of alkyl-C/*O*-alkyl-C and aromaticity increased with decomposition of litter from the Oi, Oe, to Oa horizon. The ratio of alkyl-C/*O*-alkyl-C also increased from coarse (> 250 μm) to very fine (< 2 μm) particle fractions, which indicates increased humification of soil organic matter (SOM) in the fine-sized fractions. However, aromaticity tended to decrease with decreasing particle size, so it may not be useful in evaluating SOM humification of different particle-size fractions.

Conclusions

The humification degree of the samples from O horizons and different particle-size fractions of the O/A horizon showed no gradient change with change in topography. This prevalent slow decomposition of organic matter in these perhumid climate conditions may narrow the difference in humification from the summit to lakeshore.

BACKGROUND

Humus substances are the most recalcitrant and major fraction of soil organic matter (SOM). The humus in soils has beneficial effects on plant nutrient supply, soil structure and water-holding capacity. Because of high stability, humus benefits carbon storage in soils. However, decomposition of humus releases CO_2, which may contribute to increased atmospheric CO_2 level and the greenhouse effect (Piccolo, 1996).

Chamaecyparis cypress forest is a valuable natural resource in eastern Asia for its high-quality timber. As well, the accumulation of non-decomposed organic matter on the forest floor provides a critical buffer to retain water and prevent soil erosion with heavy rainfall in montane areas. The dead timber and thick SOM might store a previously underestimated large C pool. Several decades ago, before large-scale logging, *Chamaecyparis* forest was widely distributed in cloudy montane areas in Taiwan, at about 800 to 2800 m a.s.l.

Much effort has been invested in investigating soil properties and fertility management of Japanese hinoki cypress (*Chamaecyparis obtusa*) plantations (e.g., Inagaki et al. 2008; 2011). By contrast, characteristics of humic substances under natural *Chamaecyparis* forests in this subtropical montane area are not well known because most of the preserved natural *Chamaecyparis* forests are located in steep and remote areas with poor access to roads. The Chi-Lan Mountain contains one of the few natural preserved *Chamaecyparis* forests in northcentral Taiwan.

Podzolic soils are commonly found under *Chamaecyparis* forests in the cold and humid subalpine region in Taiwan (Chiu et al. 1999; Jien et al. 2010a, b2013). The extremely high acidity and soil moisture in such undisturbed *Chamaecyparis* forest results in low diversity of soil bacterial communities (Lin et al. 20102011) as well as low decomposition of organic matter and humification degree of humic acids (Chung et al. 2012). Our previous studies revealed that the topography and intrinsic properties of different fractions of organic matter in a mountain lake environment affect the distribution and migration of organic substances (Chen and Chiu 2000). The degree of humification of SOM decreased slightly from the summit to lakeshore, and the relatively low degree of humification was due to high precipitation and acidity (Chung et al. 2012).

Particle-size fractionation has been widely used to study the physical and chemical properties and decomposition of SOM (Joliveta et al. 2006; Muñoz et al. 2009). Our previous study indicated that the humification degree of SOM increased from coarse to fine fraction size in humid subalpine soils (Chen and Chiu 2003). We hypothesized that SOM humification might be hampered differently by varied decomposition with different particle-size fractions, with possible interactions of SOM characteristics, particle-size fractions and soil moisture conditions. We aimed to clarify the differences in organic matter composition in various particle-size fractions as affected by topography in a *Chamaecyparis* forest.

METHODS

Study Site and Sampling

The Yuanyang Lake ecosystem (24°35′N, 121°24′E) is located in the northcentral part of Taiwan from 1670 to 2169 m a.s.l. This region is a cloud forest system because of annual precipitation of 4000 mm and annual mean temperature 13°C, with a temperate perhumid climate. This area was selected as a representative long-term ecological study site of Taiwan. The vegetation is dominated by Hinoki cypress (*C. obtusa*) and Taiwan false cypress (*C. formosensis*) with an understory evergreen broadleaf shrub (*Rhododendron formosanum*).

The forest soils were characterized as Albaquult, Dystrochrept and Histosol in the summit, footslope and lakeshore regions, respectively (Chiu et al. 1999; Chen and Chiu 2000). The organic matter at the summit and footslope is mor type, which is a thick mat of non-decomposed to partially decomposed O horizons, and humic peat type at the lakeshore with continuous water saturation, for a moderately decomposed class of peat characterized by 1/3 to 2/3 recognizable plant fibers. Soil at the summit is well drained, whereas that at the footslope and lakeshore is poorly drained. In particular, the lakeshore, located about 1.5 m above the lake, frequently experiences inundation of water during the monsoon season.

We selected 3 representative pedons along a topographic sequence at the summit, footslope and lakeshore. At each study site, a pit was excavated for describing macromorphological soil characteristics and for collecting soil samples according to standard procedures (Soil Survey Staff2006). The toposequence for clay mineralogical characterization in this study site has been described elsewhere (Pai et al. 2007). Organic horizons (i.e., Oi, Oe and Oa) were classified and collected on the basis of rubbed fiber content that can be identified by eyes and fingers.

In addition, we collected the O/A horizon for comparing particle-size fractions. Three replicate subsamples were collected from each topographic position with use of a soil auger 8 cm in diameter and 10 cm deep before being bulked together. Non-decomposed and partially decomposed plant litter was removed before sampling.

Physical and Chemical Analyses

Soil samples first underwent low-energy sonication and then were separated into different particle-size fractions by a combination of wet sieving and continuous flow centrifugation. Four particle-size fractions, including coarse (> 250 μm), medium (53–250 μm), fine (2–53 μm), and very fine (< 2 μm), were separated (Chen and Chiu 2003). All fractionated samples were freeze-dried and stored. Samples collected from Oi, Oe, and Oa horizons were dried at 70°C and ground for analyses.

Solid-state CP/MAS ^{13}C NMR determination involved use of a Bruker DSX 400 MHz instrument operating at 100.46 MHz and spin rate 7 kHz. The 0.5-g soil sample was used each time. Acquisition parameters were contact time, 1 ms; pulse delay time, 1 s; and spectra plotted region, 0 to 200 ppm (Chen and Chiu 2003; González-Pérez et al. 2008; Novak and Smeck 1991). About 10,000 scans were collected for samples (Quideau et al. 2001). The C functional groups were determined by the following chemical-shift areas (Novak and Smeck 1991; Chen and Chiu 2003; Jien et al. 2011): alkyl-C signals, 0 to 50 ppm; *O*-alkyl-C signals, 50 to 90 ppm; di-*O*-alkyl C signals, 90 to 110 ppm; aromatic-C signals, 110 to 165; and carboxyl-C signals, 165 to 190 ppm. The methoxyl-C of the chemical shift signal appeared at 56 ppm and overlapped with the *O*-alkyl-C signals (Knicker 2000). The signal intensities in the respective chemical shift regions are expressed as a percentage of the total spectra area. The relative contents of different chemical structures were calculated from the area under the spectra. To compare each spectrum for samples, we chose the peak of carboxyl C as the standard peak and standardized the signals by adjusting the peak area to the same size for each sample. Then, we compared the peaks of different functional groups in terms of relative percentage of each functional C group. The area for each functional C group peak was calculated by use of Topspin software for the NMR instrument.

Chemical shift regions were used to calculate the ratio of alkyl-C/*O*-alkyl-C (Baldock et al. 1997) and aromaticity (Hatcher et al. 1981; Almendros et al. 2000) as follows:

$$\text{Ratio of alkyl}-\text{C}/\text{O}-\text{alkyl}-\text{C} = \frac{\text{alkyl}-\text{C peak area}(0-50\text{ppm})}{(\text{o}-\text{alkyl}-\text{C}-\text{di}-\text{O}-\text{alkyl C})\text{peak area}(0-50\text{ppm})} \quad (1)$$

$$\text{Aromaticity} = \frac{\text{aromatic C peak area}(110-165\text{ppm})}{((\text{alkyl}-\text{C}+\text{O}-\text{alkyl}-\text{C}+\text{di}-\text{O}-\text{alkyl C}+\text{aromatic C})\text{peak area}(0-165\text{ppm}))} \quad (2)$$

The above ratios were used to evaluate the humification degree and decomposition of SOM.

Chemical Analysis

The total C (TC) and total N (TN) contents in the samples from O horizons and samples of particle-size fractions from the O/A horizon were determined after combustion in a Nitrogen Analyzer (NA1500 Series 2, Fisons, Italy).

RESULTS

Chemical Characteristics and ^{13}C NMR Analyses

The pH of the soil samples was strongly acidic (3.5 ~ 3.8) and not very different in profiles. The TC was higher at the footslope and summit than the lakeshore (Table 1). The trend for TN content was similar to that for TC content. The C/N ratio decreased with depth in the 3 horizons (Oi, Oe, and Oa): highest at Oi and lowest at Oa. In addition, the C/N ratio increased from the summit to lakeshore, particularly for the Oi horizon.

Table 1: The chemical properties of partially decomposed materials collected at various organic horizons in 3 topographic positions of the Yuan yang Lake ecosystem in Taiwan

Pedons	Soil type	Organic horizon	Depth (cm)	pH (H_2O)	TC [a] (%)	TN [b] (%)	**C/N ratio**
Summit	Albaquult	O i	8-10	3.7	35.2	1.4	25.1
		O e	3-8	3.5	49.2	2.1	23.4
		O a	0-3	3.5	49.2	2.2	22.3
Footslope	Dystrochrept	O i	11-15	3.8	54.2	1.8	30.1
		O e	3-11	3.5	51.2	2.1	24.3
		O a	0-3	3.3	47.7	2.3	20.7
Lakeshore	Histosol	O i	24-33	3.5	36.3	1.1	33.0
		O e	11-24	3.4	18.7	0.6	31.1
		O a	0-11	3.5	17.7	0.7	23.9

[a] Total carbon.

[b] Total nitrogen.

Chen *et al.*

Chen *et al. Botanical Studies* 2013 **54**:56, doi: 10.1186/1999-3110-54-56

The TC and TN content in the samples for the O/A horizon increased with decreasing particle size (Table 2). The topographic position did not affect TC and TN content in different fractions of the O/A horizon samples. The C/N ratios gradually decreased with decreasing particle size at the 3 topographic positions.

Table 2: Total C, N content and C/N ratio in each particle-size fraction of soils collected from the O/A horizon at 3 topographic positions

Topographic position	Particle-size (μm)	TC [a] (%)	TN [b] (%)	C/N
Summit	Coarse (>250)	44.4	1.1	39.0
	Medium (250 ~ 53)	43.8	1.5	29.5
	Fine (53 ~ 2)	44.7	2.0	21.9
	Very Fine (< 2)	46.9	2.4	19.6
Footslope	Coarse (>250)	41.9	1.9	22.0
	Medium (250 ~ 53)	45.8	2.0	22.4
	Fine (53 ~ 2)	38.4	1.8	21.8
	Very Fine (< 2)	47.3	2.3	20.7
Lakeshore	Coarse (>250)	42.0	1.5	27.7
	Medium (250 ~ 53)	47.0	1.7	28.1
	Fine (53 ~ 2)	49.1	1.9	26.5
	Very Fine (< 2)	47.5	1.8	26.5

[a]Total carbon.

[b]Total nitrogen.

Chen *et al.*

Chen *et al. Botanical Studies* 2013 **54**:56, doi: 10.1186/1999-3110-54-56

Figure 1 shows the characteristics of CP/MAS ^{13}C NMR spectra of partially decomposed plant materials at the 3 horizons. The relative distribution of integrated peak areas of chemical shift regions in the CP/MAS ^{13}C NMR for different carbon functional groups of organic matter from each horizon was in the order of alkyl-C > *O*-alkyl-C > aromatic-C > di-*O*-alkyl-C > carboxyl-C at all 3 topographical positions (Table 3). The content of alkyl-C and carboxyl-C increased with depth at the footslope and lakeshore but not at the summit. By contrast, the content of *O*-alkyl-C and di-*O*-alkyl-C slightly decreased with depth, but that of aromatic-C and carboxyl-C slightly increased with depth at all positions.

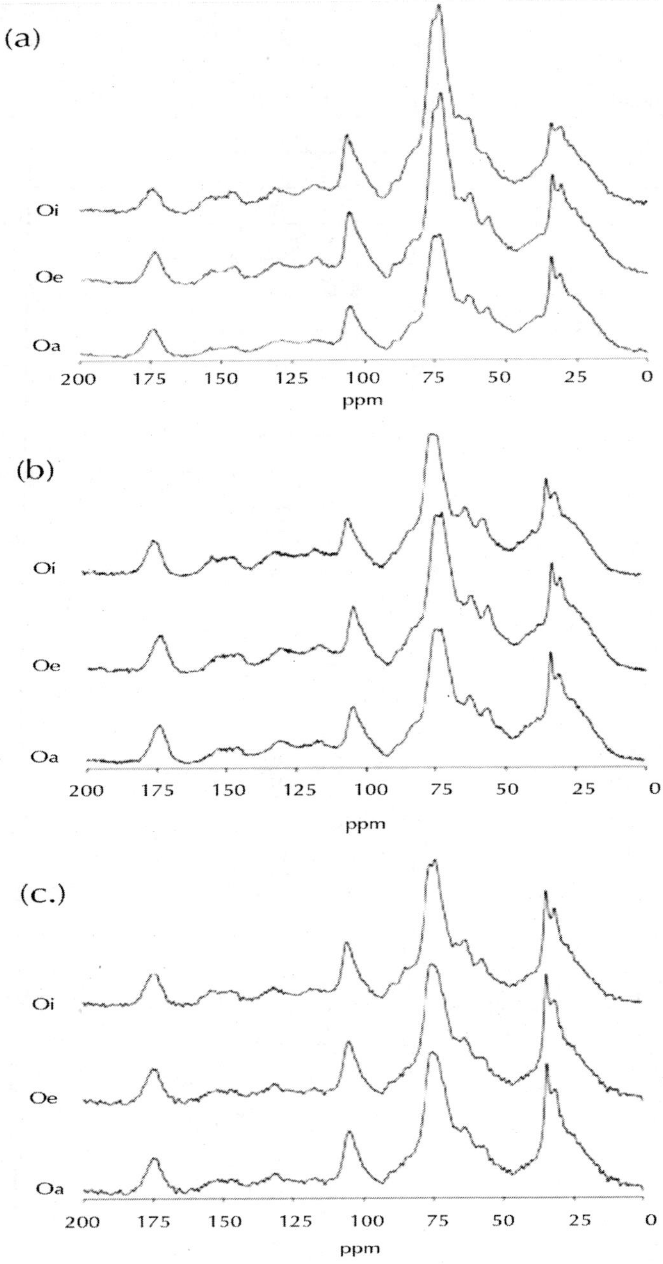

Figure 1: CP/MAS ^{13}C NMR spectra for partially decomposed plant materials in 3 organic horizons (O i, O e, and O a) at the (a) summit, (b) foot slope, and (c) lakeshore

Table 3: Relative distribution of integrated peak areas of chemical shift regions in the CP/MAS ^{13}C NMR spectra of partially decomposed materials collected from 3 organic horizons at 3 topographic positions

Topographic position	**Horizon**	**C functional group (%)**					**A/ *O*-A ratio**[a]	Aromaticity[b]
		alkyl	*O*-alkyl	di- *O* -alkyl	aromatic	carboxyl		
Summit	Oi	34.1	30.5	15.0	13.7	6.7	0.75	0.15
	Oe	33.7	30.5	13.5	16.4	5.9	0.77	0.17
	Oa	33.5	30.0	12.5	16.2	7.8	0.79	0.18
Footslope	Oi	26.7	31.6	15.4	18.3	8.0	0.57	0.20
	Oe	28.2	30.9	14.9	18.2	7.8	0.61	0.20
	Oa	27.8	30.1	14.4	19.9	7.7	0.62	0.22
Lakeshore	Oi	30.9	26.4	15.4	19.9	7.4	0.74	0.21
	Oe	31.0	26.1	14.6	20.4	7.8	0.76	0.22
	Oa	31.3	25.9	14.2	20.7	7.9	0.78	0.22

^{13}C NMR Analyses of Particle-Size Fractions of O/A Horizon

Coarse (> 250 μm) particle size was the dominant (61.0 ~ 83.3%) fraction, followed by medium (250–53 μm) and fine fractions (53–2 μm), with the minimum (1.1 ~ 2.1%) being the very fine fraction (< 2 μm) (Table 4). The ^{13}C NMR spectra for organomineral (O/A horizon) particle-size fractions (Figure 2) and the results of peak area integrations (Table 5) from the summit to lakeshore demonstrated that fraction size largely affected the C structure. The distribution of C functional groups showed a predominance of *O*-alkyl-C and alkyl-C groups and a relatively low content of aromatic-C groups in the samples. The amount of *O*-alkyl-C, di-*O*-alkyl-C and aromatic-C decreased with decreasing particle size, and alkyl-C content increased with decreasing particle size. The aromatic-C content was lower with fine than coarse particle size. The alkyl-C/*O*-alkyl-C ratio and aromaticity responded to particle size differently, with an increase for alkyl-C/*O*-alkyl-C ratio and a decrease in aromaticity from coarse to fine particle size. In comparing different topographic positions, *O*-alkyl-C and di-*O*-alkyl-C contents were highest at the lakeshore, particularly in coarse and medium fractions (Table 5). By comparison, alkyl-C content was lowest at the lakeshore, particularly in the finest particle-size fraction.

Table 4: Particle-size distribution of soils collected from the O/A horizon (0–10 cm) at 3 topographic positions

Topographic	**Particle size (μm)**			
position	Very fine (< 2)	Fine (2–53)	Medium (53–250)	Coarse (>250)
Summit	1.6	19.8	17.5	61.0
Foot slope	2.1	15.1	18.2	64.5
Lakeshore	1.1	6.0	9.5	83.3

Data are percentages.

Chen et al.

Chen et al. Botanical Studies 2013 54:56, doi: 10.1186/1999-3110-54-56

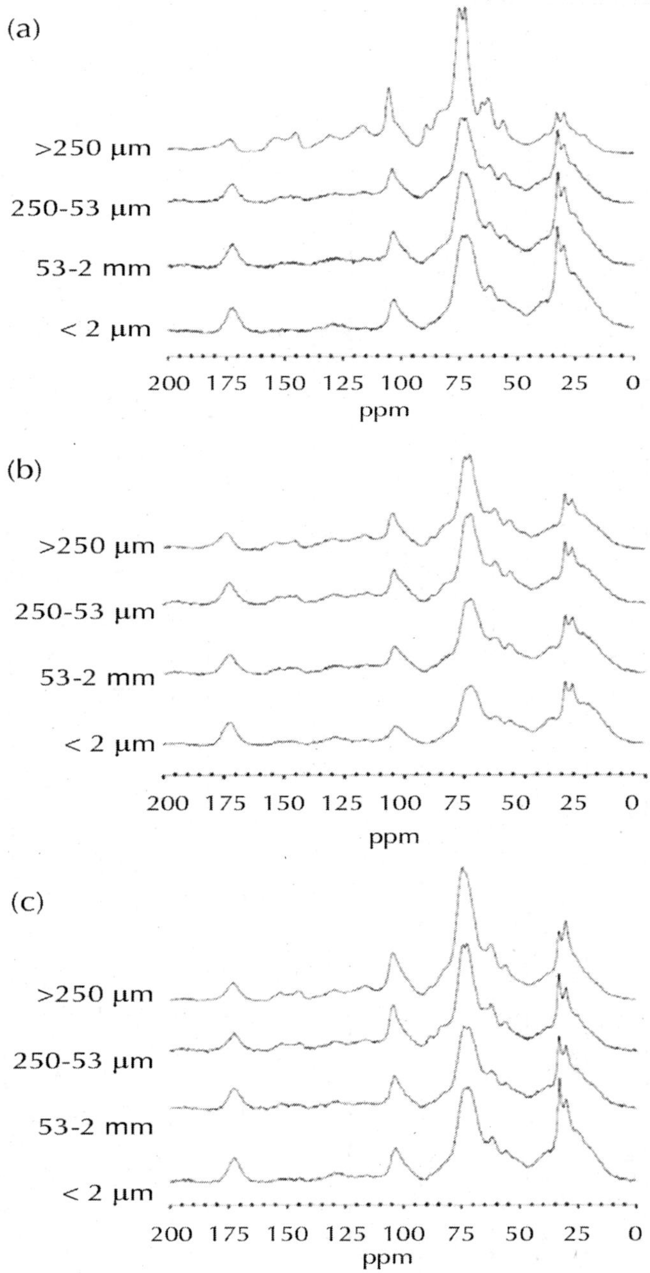

Figure 2: CP/MAS ^{13}C NMR spectra for soil organic matter in different particle-size fractions from the O/A horizon at the (a) summit, (b) footslope, and (c) lakeshore.

Table 5: Relative intensities of the CP/MAS ^{13}C NMR spectra in different particle-size fractions isolated from the O/A horizon (0–10 cm) in 3 topographic positions

Topographic	Particle-size	C functional groups (%)					A/ *O* -A ratio [a]	Aromaticity [b]
position	**(μm)**	**alkyl**	***O*-alkyl**	**di- *O* -alkyl**	**aromatic**	**carboxyl**		
Summit	Coarse (>250)	28.4	42.9	9.9	13.7	5.1	0.54	0.14
	Medium (250 ~ 53)	31.2	42.4	9.3	11.8	5.3	0.60	0.12
	Fine (53 ~ 2)	36.9	41.8	8.2	8.3	4.7	0.74	0.09
	Very fine (<2)	41.2	41.7	7.1	5.3	4.6	0.84	0.06
Foot slope	Coarse (>250)	29.5	41.5	9.5	13.5	6.0	0.58	0.14
	Medium (250 ~ 53)	31.8	42.2	8.9	11.5	5.5	0.62	0.12
	Fine (53 ~ 2)	34.9	39.4	8.5	11.4	5.8	0.73	0.12
	Very fine (<2)	41.5	35.7	6.8	9.3	6.7	0.98	0.10
Lakeshore	Coarse (>250)	28.8	45.3	10.2	11.1	4.6	0.52	0.12
	Medium (250 ~ 53)	28.4	44.4	10.6	11.8	4.9	0.52	0.12
	Fine (53 ~ 2)	34.7	41.8	8.7	9.2	5.6	0.69	0.10
	Very fine (<2)	38.3	42.1	7.8	6.9	4.9	0.77	0.07

[a] alkyl-C peak area (0–50 ppm)/((*O*-alkyl-C + di-*O*-alkyl C) peak area (50–110 ppm)).

[b] aromatic C peak area (110–165 ppm)/ ((alkyl-C + *O*-alkyl-C + di-*O*-alkyl C + aromatic C) peak area (0–165 ppm)).

Chen *et al.*

Chen *et al.* *Botanical Studies* 2013 54:56, doi: 10.1186/1999-3110-54-56

DISCUSSION

Chemical Composition

Samples from 3 organic horizons of the Yuanyang Lake ecosystem predominantly consisted of organic matter. The high content of C in these horizons was consistent with that in other perhumid forests (Mafra et al. 2007; Schawe et al. 2007). Poor drainage and flooding in riparian soils diminishes the decomposition of SOM but improves the denitrification by microorganisms (Mafra et al. 2007). Thus, we found lower TN content and higher C/N ratio at the lakeshore than footslope and summits (Table 1). The C/N ratio varied markedly by particle-size fraction at the summit but not greatly at the footslope and lakeshore (Table 2). High total organic C content throughout the fractions indicates the accumulation of organic matter in the soil. In addition, high percentages (61% to 83%) of samples were in the coarse fraction with decomposing plant residues (Table 4), which suggests the retardation of organic matter decomposition in such perhumid climate conditions.

Carbon Functional Groups in Different Particle-Size Fractions

We found *O*-alkyl-C and di-*O*-alkyl-C as the dominant components in the entire soil fraction, which can probably be used by microorganisms during humification processes (Keeler et al. 2006). The region of aromatic-C includes phenolic-C, derived from lignin and tannin, and bacterial resynthesized compounds consisting of alkyl-C and carboxyl-C (Mahieu et al. 1999; Mathers et al.2000; Ussiri and Johnson 2003; López et al. 2008).

Physical fractionation of soil according to particle size followed by chemical, biological, and physical analyses of fractions is a powerful tool in process-oriented SOM research (Mao et al. 2007). Particle-size fractioning allows for separating SOM pools of varying degrees of microbial alteration and mineral association (Joliveta et al. 2006). The SOM in the coarse fraction primarily consisted of labile plant residues, whereas microbial biomass is supposed to be concentrated in the very fine fraction (Zech et al. 1996; Kimetu et al. 2008). Thus, SOM associated with very fine fractions tends to be more aliphatic than does whole SOM (Mao et al. 2007; Jiménez et al. 2008; Muñoz ET al.2009). We found low C/N ratio (Table 2) and high alkyl-C content (Table 5) in very fine particle-size fractions, which suggests the loss of easily decomposable carbohydrates and selective preservation of inherently recalcitrant materials during the plant residue decomposition (Mathers et al. 2000; Wagai et al. 2008; Rovira et al. 2009).

The *O*-alkyl-C content in coarse and medium fractions was higher at the lakeshore than at the summit and footslope. In particular, peaks at 56, 63, 84 and 89 ppm (*O*-alkyl-C) were completely diminished from coarse (>250 μm) to very fine (<2 μm) at the summit, with only a trace amount at the footslope and lakeshore (Figure 2). The *O*-alkyl-C content might be greater at the lakeshore than other positions and was universally lower in fine than coarse particles (Table 5), as was found in our previous study of subalpine forest and grassland soils (Chen and Chiu 2003). Thus, the spectra of the coarse fraction (>250 μm) resembled that of decomposing materials (Figures 1 and2), which resulted from most of the soils being in a coarse fraction (Table 4).

Humification Degree of Organic Horizons and Particle-Size Fractions of the O/A Horizon

The humification degree and characteristics of SOM could be evaluated by ratio of alkyl-C/*O*-alkyl-C and aromaticity. The high ratio of alkyl-C/*O*-alkyl-C and aromaticity indicates the high humification degree of SOM in the soils (Baldock et al. 1997; Almendros et al. 2000; Chen et al.2004; Mueller and Koegel-Knabner 2009). The alkyl-C/O-alkyl-C ratios of the Oi, Oe, and Oa horizons in the 3 positions increased with increasing soil depth (Table 3), which indicates a higher humification degree in the Oa horizon. We found no typical pattern for alkyl-C/O-alkyl-C ratios in O horizons along topographical positions (Table 3), which indicates restrained decomposition and humification because of low temperature and high annual precipitation. The results for aromaticity were similar to those for alkyl-C/O-alkyl-C (Table 3), which could reflect the humification state of SOM. The influence of litter input on changes in remnant masses of each carbon during the early humification processes showed that mass loss rate of aromatic C in humified litter is higher in cypress than cedar (Ono et al. 20092011). Our previous studies showed that humification degree of HAs, determined by E4/E6 ratio (Chen et al. 2001a, b) and by ^{13}C NMR (Chung et al. 2012) differed little by topographic position of the study site.

The ratio of alkyl-C/*O*-alkyl-C increased with decreasing particle-size fraction (Table 5), so high stability in the fine fraction is due to the protection by clay minerals (Wagai et al. 2008) and to the chemical resistance of the alkyl-C structure to decomposition. The humification degree in the medium and fine fractions of the O/A horizon samples was similar to that in the O horizons in the 3 studied positions, which reflects the chemical properties in SOM being strongly dominated by the distribution of particle-size fractions. The ratios of alkyl-C/*O*-alkyl-C in each particle-size fraction with topographic change under the *Chamaecyparis* forest were similar in value and variation and were lower than in *Tsuga* forest soils (Chen and Chiu 2003). Thus, the relatively low proportion of alkyl-C/*O*-alkyl-C ratios is due to the poor decomposition under perhumid climate conditions.

By comparison, aromaticity findings were opposite to those for ratios of alkyl-C/*O*-alkyl-C in particle-size fraction. Aromaticity decreased with decreasing particle size (Table 5). The present results agree with the suggestions of Hempfling et al. (1987) and Almendros et al. (2000 () questioning the increase in aromaticity during humification in soils. The results of Baldock et al.1997) also showed that aromatic C contents tended to decrease with decreasing particle size, which suggests that increased extent of decomposition was not accompanied by an increase in aromatic C content. If aromatic C content did not accumulate with decomposition, an increase in aromaticity should not be used as an indicator of the extent of decomposition.

CONCLUSIONS

The distribution of particle-size fractions in soil is a critical factor in determining the chemical composition of SOM. ^{13}C NMR analysis of soil particle-size fractions revealed that the C structure of the SOM changes with different particle-size fraction. The accumulation of recalcitrant C compounds in the fine particle-size fraction was contributed by alkyl-C rather than aromatic-C. The ratio of alkyl-C/*O*-alkyl-C indicates increasing SOM humification with decreasing particle size. The humification degree showed no gradient change pattern in different topographical positions. The effect of topography on the decomposition and humification of organic matter in the*Chamaecyparis* forest was apparently overshadowed by the slow decomposition of organic matter.

AUTHORS' CONTRIBUTIONS

JSC designed and conducted the research. TLC conducted some laboratory tests. GT helped in analyzing and interpreting data. JSC and CYC wrote the manuscript with inputs from other authors. All authors read and approved the final manuscript.

ACKNOWLEDGEMENTS

This work was supported in part by the National Science Council, Taiwan (NSC 101-2621-B-001-002-MY3).

REFERENCES

1. Almendros G, Dorado J, González-Vila FJ, Blanco MJ, Lankes U (2000) ^{13}C NMR assessment of decomposition patterns during composting of forest and shrub biomass. Soil Biol Biochem 32:793-804.
2. Baldock JA, Oades JM, Nelson PN, Skene TM, Golchin A, Clarke P (1997) Assessing the extent of decomposition of natural organic materials using solid-state ^{13}C NMR spectroscopy. Aust J Soil Res 35:1061-1083
3. Chen JS, Chiu CY (2000) Effect of topography on the composition of soil organic substances in a perhumid sub-tropical montane forest ecosystem in Taiwan. Geoderma 96:19-30
4. Chen JS, Chiu CY, Nagatsuka S (2001) Spectral features of humic substances in a perhumid subtropical montane forest ecosystem, Taiwan. Soil Sci Plant Nutr 47:179-185

5. Chen JS, Chiu CY (2003) Characterization of soil organic matter in different particle-size fractions in humid subalpine soils by CP/MAS C^{13} NMR. Geoderma 117:129-141
6. Chen MC, Wang MK, Chiu CY, Huang PM, King HB (2001) Determination of low molecular weight dicarboxylic acids and organic functional groups in rhizosphere and bulk soils of Tsuga and Yushania in a temperate rain forest. Plant Soil 231:37-44
7. Chen CR, Hu ZH, Mathers NJ (2004) Soil carbon pools in adjacent natural and plantation forests of subtropical Australia. Soil Sci Soc Am J 68:282-291
8. Chiu CY, Lai SY, Lin YM, Chiang HC (1999) Distribution of the radionuclide Cs-137 in the soils of a wet mountainous forest in Taiwan. Appl Radiat Isot 50:1097-1103
9. Chung TL, Chen JS, Chiu CY, Tian G (2012) ^{13}C-NMR spectroscopy studies of humic substances in subtropical perhumid montane forest soil. J For Res 17:458-467
10. González-Pérez M, Torrado PV, Colnago LA, Martin-Neto L, Otero XL, Milori DMBP, Gomes FH (2008) ^{13}C NMR and FTIR spectroscopy characterization of humic acids in spodosols under tropical rain forest in southeastern Brazil. Geoderma 146:425-433
11. Hatcher PG, Schnitzer M, Dennis LW, Maciel GE (1981) Aromaticity of humic substance in soils. Soil Sci Soc Am J 45:1089-1094
12. Hempfling R, Ziegler F, Zech W, Schulten HR (1987) Litter decomposition and humification in acidic forest soils studied by chemical degradation, IR and NMR spectroscopy and pyrolysis field ionization mass spectrometry. Zeitschrift für Pflanzenernährung und Bodenkunde 150:179-186
13. Inagaki Y, Kuramoto S, Torii A, Shinomiya Y, Fukata H (2008) Effects of thinning on leaf-fall and leaf-litter nitrogen concentration in hinoki cypress (*Chamaecyparis obtusa* Endlicher) plantation stands in Japan. For Ecol Manag 255:1859-1867
14. Inagaki Y, Nakanishi A, Fukata H (2011) Soil properties and nitrogen utilization of hinoki cypress as affected by strong thinning under different climatic conditions in the Shikoku and Kinki districts in Japan. J For Res 16:405-413

15. Jiménez JJ, Lal R, Russo RO, Leblanc HA (2008) the soil organic carbon in particle-size separates under different regrowth forest stands of north eastern Costa Rica. Ecol Eng 34:300-310
16. Jien SH, Wu SP, Chen ZS, Chen TH, Chiu CY (2010) Characteristics and pedogenesis of podzolic forest soils along a toposequence near a subalpine lake in northern Taiwan. Bot Stud 51:223-236
17. Jien SH, Hseu ZY, Iizuka Y, Chen TH, Chiu CY (2010) geochemical characterization of placic horizons in subtropical montane forest soils, Northeastern Taiwan. Eur J Soil Sci 61:319-332
18. Jien SH, Chen TH, Chiu CY (2011) Effects of afforestation on soil organic matter characteristics under subtropical forests with low elevation. J For Res 16:275-283
19. Jien SH, Pai CW, Iizuka Y, Chiu CY (2013) Pedogenic processes of placic and spodic horizons in sub-tropical sub-alpine forest soils with contrasting textures. Eur J Soil Sci 64:423-434
20. Joliveta C, Angers DA, Chantigny MH, Andreux F, Arrouays D (2006) Carbohydrate dynamics in particle-size fractions of sandy spodosols following forest conversion to maize cropping. Soil Biol Biochem 38:2834-2842
21. Keeler C, Kelly EF, Maciel GE (2006) Chemical-structural information from solid-state C-13 NMR studies of a suite of humic materials from a lower montane forest soil, Colorado, USA. Geoderma 130:124-140
22. Knicker H (2000) Biogenic nitrogen in soils as revealed by solid-state carbon-13 and nitrogen-15 nuclear magnetic resonance spectroscopy. J Environ Qual 29:715-723
23. Kimetu JM, Lehmann J, Ngoze S, Mugendi DN, Kinyangi JM, Riha S, Verchot L, Recha JW, Pell A (2008) Reversibility of soil productivity decline with organic matter of differing quality along a degradation gradient. Ecosystems 11:726-739
24. Lin YT, Huang YJ, Tang SL, Whitman WB, Coleman DC, Chiu CY (2010) Bacterial community diversity in undisturbed perhumid montane forest soils in Taiwan. Microb Ecol 59:369-378.
25. Lin YT, Jangid K, Whitman WB, Coleman DC, Chiu CY (2011) Soil bacterial communities in native and regenerated perhumid montane forests. Appl Soil Ecol 47:111-118

26. López R, Gondar D, Iglesias A, Fiol S, Antelo J, Arce F (2008) Acid properties of fulvic and humic acids isolated from two acid forest soils under different vegetation cover and soil depth. Eur J Soil Sci 59:892-899

27. Mafra AL, Senesi N, Brunetti G, Miklós AAW, Melfi AJ (2007) Humic acids from hydromorphic soils of the upper Negro river basin, Amazonas: Chemical and spectroscopic characterization. Geoderma 138:170-176

28. Mahieu N, Powlson DS, Randall EW (1999) Statistical analysis of published carbon-13 CPMAS NMR spectra of soil organic matter. Soil Sci Soc Am J 63:307-319

29. Mao J, Fang X, Schmidt-Rohr K, Carmo AM, Hundal LS, Thompson ML (2007) Molecular-scale heterogeneity of humic acid in particle-size fractions of two Iowa soils. Geoderma 140:17-29

30. Mathers NJ, Mao XA, Saffigna PG, Xu ZH, Berners-Price SJ, Perera MCS (2000) Recent advances in the application of ^{13}C and ^{15}N NMR spectroscopy to soil organic matter studies. Aust J Soil Res 38:769-787

31. Mueller CW, Koegel-Knabner I (2009) Soil organic carbon stocks, distribution, and composition affected by historic land use changes on adjacent sites. Biol Fertil Soils 45:347-359

32. Muñoz C, Monreal C, Schnitzer M, Zagal E (2009) Analysis of soil humic acids in particle-size fractions of an alfisol from a Mediterranean-type climate. Geoderma 151:199-203

33. Novak JM, Smeck NE (1991) Comparisons of humic substances extracted from contiguous alfisols and mollisols of southwestern Ohio. Soil Sci Soc Am J 55:96-102

34. Ono K, Hirai K, Morita S, Ohse K, Hiradate S (2009) Organic carbon accumulation processes on a forest floor during an early humification stage in a temperate deciduous forest in Japan: Evaluations of chemical compositional changes by ^{13}C NMR and their decomposition rates from litterbag experiment. Geoderma 151:351-356

35. Ono K, Hiradate S, Morita S, Ohse K, Hirai K (2011) Humification processes of needle litters on forest floors in Japanese cedar (*Cryptomeria japonica*) and Hinoki cypress (*Chamaecyparis obtusa*) plantations in Japan. Plant Soil 338:171-181

36. Pai CW, Wang MK, Chiu CY (2007) Clay mineralogical characterization of a toposequence of perhumid subalpine forest soils in northeastern Taiwan. Geoderma 138:177-184
37. Piccolo A (1996) Humus and soil conservation. In: Piccolo A (ed) Humic Substances in Terrestrial Ecosystems, Elsevier, Amsterdam. pp 225-264
38. Rovira P, Duguy B, Vallejo VR (2009) Black carbon in wildfire-affected shrubland Mediterranean soils. J Plant Nutr Soil Sci 172:43-52
39. Quideau SA, Chadwick OA, Benesi A, Graham RC, Anderson MA (2001) A direct link between forest vegetation type and soil organic matter composition. Geoderma 104:41-60
40. Schawe M, Glatzel S, Gerold G (2007) Soil development along an altitudinal transect in a Bolivian tropical montane rainforest: Podzolization vs. hydromorphy. Catena 69:83-90
41. Soil Survey Staff (2006) Keys to Soil Taxonomy. United States Department of Agriculture, Washington DC.
42. Ussiri DAN, Johnson CE (2003) Characterization of organic matter in a northern hardwood forest soil by ^{13}C NMR spectroscopy and chemical methods. Geoderma 111:123-149
43. Wagai R, Mayer LM, Kitayama K, Knicker H (2008) Climate and parent material controls on organic matter storage in surface soils: A three-pool, density-separation approach. Geoderma 147:23-33
44. Zech W, Guggenberger G, Haumaier L, Pöhhacker R, Schäfer D, Amelung W, Miltner A, Kaiser K, Ziegler F (1996) Organic matter dynamics in forest soils of temperate and tropical ecosystems. In: Humic Substances in Terrestrial Ecosystems. Elsevier Science, B.V., Amsterdam. pp 101-170

Citations

CHAPTER 1

Silvia Vaccaro, Andrea Ertani, Antonio Nebbioso, Adele Muscolo, Silvia Quaggiotti, Alessandro Piccolo, and Serenella Nardi, Humic substances stimulate maize nitrogen assimilation and amino acid metabolism at physiological and molecular level, doi: 10.1186/s40538-015-0033-5.

CHAPTER 2

Catello Pane, Giuseppe Celano, Alessandro Piccolo, Domenica Villecco, Riccardo Spaccini, Assunta M Palese, and Massimo Zaccardelli, Effects of on-farm composted tomato residues on soil biological activity and yields in a tomato cropping system, doi: 10.1186/s40538-014-0026-9.

CHAPTER 3

Jie Li, Yanli Yi, Xilei Cheng, Dageng Zhang, and Muhammad Irfan, Study on the effect of magnetic field treatment of newly isolated Paenibacillus sp, doi:10.1186/s40529-015-0083-9.

CHAPTER 4

Wendy Peterman and Richard H Waring, Does overshoot in leaf development of ponderosa pine in wet years leads to bark beetle outbreaks on fine-textured soils in drier years?, doi:10.1186/s40663-014-0024-1.

CHAPTER 5

Qizheng Mao, Ganlin Huang, Alexander Buyantuev, Jianguo Wu, Shanghua Luo, and Keming Ma, Spatial Heterogeneity of Urban Soils: the Case of the Beijing Metropolitan Region, China, doi:10.1186/s13717-014-0023-8.

CHAPTER 6

Rodrigo Tizón, Juan Pedro Wulff, and Daniel Valerio Peláez, The effect of increase in the temperature on the foraging ofAcromyrmex lobicornis (Hymenoptera: Formicidae), doi: 10.1186/s40555-014-0040-4.

CHAPTER 7

Gergely Tóth, Ciro Gardi, Katalin Bódis, Éva Ivits, Ece Aksoy, Arwyn Jones, Simon Jeffrey, Thorum Petursdottir, and Luca Montanarella, Continental-Scale Assessment of Provisioning Soil Functions In Europe, doi:10.1186/2192-1709-2-32.

CHAPTER 8

Jenn-Shing Chen, Tay-Lung Chung, Guanglong Tian, and Chih-Yu Chiu, Characterization of Soil Organic Matter in Perhumid Natural Cypress Forest: Comparison of Humification in Different Particle-Size Fractions, doi:10.1186/1999-3110-54-56

Index

U

V

W

Y